Gene Worship

Gene Worship

Moving Beyond the Nature/Nurture Debate over Genes, Brain, and Gender

Gisela Kaplan
and
Lesley J. Rogers

Other Press
New York

First softcover printing 2010
ISBN 978-1-59051-443-6

Production Editor: Robert D. Hack

This book was set in 11 pt. Goudy by Alpha Graphics of Pittsfield, NH.

 Printed in the United States of America on acid-free paper. For information write to Other Press LLC, 2 Park Avenue, 24th Floor, New York, NY 10016. Or visit our Web site: www.otherpress.com.

Library of Congress Cataloging-in-Publication Data

Kaplan, Gisela.
Gene worship : moving beyond the nature/nurture debate over genes, brain, and gender / Gisela Kaplan and Lesley Rogers.
p. cm.
Includes bibliographical references and index.
ISBN 1-59051-034-8 (hardcover)
1. Sex differences. 2. Sex differences (Psychology). 3. Genetic psychology. 4. Nature and nurture. 5. Neurogenetics. I. Rogers, Lesley. II. Title.

QP81.5 .K37 2003
155—dc21

2002030845

Angel of History

His face is turned towards the past. Where we perceive a chain of events, he sees one single catastrophe, which keeps piling wreckage upon wreckage and hurls it in front of his feet. The angel would like to stay, awaken the dead, and make whole what has been smashed. But a storm is blowing from Paradise; it has got caught in his wings with such violence that the angel can no longer close them. The storm irresistibly propels him into the future to which his back is turned, while the pile of debris before him grows skyward. The storm is what we call progress.

Walter Benjamin, *Illuminations*, written in the late 1920s

Contents

Acknowledgments ix

Part I: In Theory

1 Freedom from Genes 3

2 Essentialisms, Determinisms 27

3 Limits of Biological Explanations 47

4 Biology and Flexibility 63

Part II: In Realities

5 Gendered World 83

6 Women's Biology and the Consequences 109

Part III: In Perpetuity

7 Designated Sex Development 141

8 Attractions 163

9 Gay or “Queer” Gene 193

10 Complexity, Plasticity 225

References 237

Index 279

Acknowledgments

We wish to express our gratitude to Judith Feher Gurewich for inviting us to write this book and our thanks to the entire team at Other Press for a very enjoyable and constructive collaboration, particularly Bob Hack for working so closely with us on the text. We also wish to thank Prof. Iréne Matthis and Prof. Mark Solms for inviting us to deliver talks in Stockholm in an intellectual environment and on topics directly relevant to this book (joint conferences on Sex and Gender, by IPA, Committee on Women and Psychoanalysis, followed by Neuroscientific Perspectives organized by the International Neuro-Psychoanalysis Society). It proved to be a stimulating and important sounding board for our ideas and reconfirmed why it is so important at this time to debate issues discussed in the book. We would also like to acknowledge very valuable feedback that we received on the draft of our manuscript from Prof. Hilary Rose, Susan Fairfield and Michael Moskowitz of Other Press, and Dr. Lynda Birke for commenting on the entire manuscript.

Acknowledgments

PART ONE

In Theory

1

Freedom from Genes

Genetic explanations for human behavior are popular and pervasive but not because they have scientific validity. Over the past decade we have heard widely promoted claims about discovery of the gene "for language"; the "gay gene" or sequence of genes; a gene that causes schizophrenia, or at least increases our chance of becoming schizophrenic; and another causing depression; together with claims of finding genes that cause Alzheimer's disease, dyslexia, alcoholism, and even homelessness. Many more medical and social syndromes have been singled out for such biological scrutiny, as if they were clearly defined units of behavior neatly matched and determined by discrete and direct action of the genes. This groundswell of gene-based explanations, stretching well beyond scientific facts, has refueled much older debates, especially about genes and sex differences in behavior. While claims about finding the existence of genes causing depression, homosexuality and so on have come and gone with the flow of claimed scientific proof and subsequent falsification, genetic explanations of a wide range of culturally fostered sex differences have persisted, often finding new and vocal supporters among the scientific community even in the face of little evidence.

Here we step aside from the gathering tide of genetic claims for causes of human behavior and show that these are simplistic explanations based on social ideas hidden falsely in scientific garb. We deal with the pitfalls and dangers of the relatively new fields of sociobiology and evolutionary psychology, by no means insignificant as they are represented by a rapidly growing number of followers. Gene technology, with its associated genetic explanations that resolve complex human behavior into simple units, has, we believe, moved us dangerously close to a new false religion, that of worshiping the genes as the key to understanding human psychology, society, and existence. At the same time as tracing the origins of these ideas, and seeing some fearsome social changes that might flow from them, we offer an optimistic antidote to their gloomy, deterministic perspective. We do so by proposing a strong case for an alternative interpretation of research on genes, brain, and behavior and we do so primarily in the age-old battleground of the sexes and sexual orientation. What really does cause the differences between men and women? What does the "gay gene" debate signify? How much outrage should we feel when contemporary scientists either inadvertently or quite openly promote a new form of eugenics?

We tackle these questions and put them in a new perspective, not by denying any role of the genes but by recognizing the complex dynamics between gene expression and environmental influences.

Discussion of sex differences raises undiluted passions and multifarious viewpoints, still as vibrant today as they have always been. But have the arguments changed with time? Have biological explanations always been used to uphold traditional sex roles, providing an apparently scientific underpinning for inequalities between women and men?

Research on differences between the sexes is bedeviled by a long history of prejudiced ideas and assumptions about causation. Old ideas about the claimed naturalness, and therefore inevitability, of sex differences in behavior, as we will show, have swayed research design and interpretation of results in every area of research investigating sex differences. Time and time again, genetic and hormonal explanations are accepted as being the fact of the matter, with only passing consideration of the potential influences of experience and learning. The last three decades have seen critiques of biological determinist theories of sex differences by feminists conversant with biology (see Barr and Birke 1998, Fausto-Sterling 1992, Keller 1985, Rose 1983, 1994). Feminists and others have tackled the mainstay of traditional biological thought by pointing out the flexibility of the biological organism and the complexity of the interaction between genes and experience during development. As Oyama (1985, 1993) has pointed out, development is not merely a process involving a battle between nature (genes) and nurture (experience) but the interweaving of dynamic processes within a system that is inseparably both the organism and its environment. These new conceptualizations, and critiques of biology, have been influential but, we regret, not influential enough to become center stage within the discipline and unable to stem the flow to popularity of genetic determinist explanations for human behavior.

We decided to write this book because we believe that it is now timelier than ever to address the issue of biological explanations of sex/gender differences, when new techniques in molecular genetics have turned the attention of a large group of biological scientists to seek genetic explanations for human behavior. A counterforce to this trend might be found in a branch of neuroscience dealing with the processes of development of organisms (often referred to as neurobiology and developmental neuroscience). Contrary to the emphasis on genetic determinism, it has revealed the plasticity of nerve cells and the brain by demonstrating their ability to change in response to a wide variety of inputs, not only during development but also, far beyond expectations, in adulthood. Although not directly investigating human behavior, this field of research presents a model of continuous change and flexibility that pertains to a very different model for understanding the associations between biology and behavior.

We can, therefore, find two rather different viewpoints in biology, but today the genetic and inflexible model for human behavior seems to be heard more often and widely, especially in the media. Strongly popularized, the voices come mainly from molecular geneticists and evolutionary psychologists. Molecular genetics is not inherently conservative, but its promise has been extrapolated beyond the laboratory and has spawned a rash of research claiming to uncover genetic causes for the behavior of humans (e.g., see Duster 1990 and Thompson et al. 2001). It has given impetus to evolutionary psychology, a field in which genetic explanations for human behavior are hypothesized, in most cases without any scientific underpinning, and based on assumptions of how ancestral humans lived (e.g., Palmer and Palmer 2002, who title their book *Evolutionary Psychology: The Ultimate Origins of Human Behavior* as if they have a new truth about our species).

Despite its youthful appearance, evolutionary psychology represents perhaps the most conservative force in biology and social life today. It rose to prominence, then largely referred to as behavioral genetics, by rekindling of the racially and gender based IQ debates of the early 1970s (i.e., those with Hans Eysenck [1971] and Arthur Jensen [1969] in the lead). These racially argued IQ debates, well rebuffed by protests and academic debate (e.g., Kamin 1977), were laid to rest for a time when "Asians" proved to have higher mean IQs than "whites" and when the now infamous twin studies of Cyril Burt, claiming to show inheritance of IQ in identical twins, were found to be fabricated. But genetic explanations for intelligence were revived again in the mid-1990s (Rushton 1994).

Evolutionary psychology arose as an offshoot of sociobiology and that, in turn, came from the fields of evolutionary biology and behavioral genetics (for a brief discussion of some of the debates see Cartwright 2001 and Ruse 2001). Disconcertingly, evolutionary psychology adheres to the narrow gene-centered ideas of evolution, and adaptation of organisms, without much caring to listen to theoretical advances made by evolutionary biologists over the last two decades or so. Evolutionary biologists recognize an important role of the environment on the expression of genes. They see genetic and developmental, or epigenetic, constraints as a reality. Exactly what these epigenetic rules are may be vague, and the balance (an important one) struck between the genetic and epigenetic factors is hotly debated, and rightly so, but the thinking of evolutionary biologists is far more sophisticated than that of evolutionary psychologists. As Pigliucci and Kaplan (2000) point out, many of the sociobiologists have now embraced this new conceptual framework allowing some consideration of influences from the environment (e.g., Alcock 2001, Wilson 1998). In other words, they have moved away from their original simplistic ideas of natural selection acting directly on the genes to consider the genetic expression as being channeled by a series of constraints, or epigenetic rules. In our view, even this model of genes and epigenetic rules still fails to recognize the dynamic flexibility of developmental processes (see Rose 1997). At least some moves are being made by sociobiologists to take "the environment" into account, and we recognize this as a distinct advance on the earlier theoretical underpinnings of sociobiology. Yet these small moves away from seeing genes as the single most powerful arbitrator of behavior have barely been embraced by evolutionary psychology, which "seems to be retracing its ancestor's [sociobiologists] path of mistakes by making wild claims on the genetic basis of human behavior while ignoring the two decades of debate" (Pigliucci and Kaplan 2000, p. 67). It might seem surprising, therefore, that the hypothesized genetic explanations for human behavior put forward by evolutionary psychologists receive such attention and popularity, but we seem to be partial to simple, and rigid, explanations for our behavior. At least, we grasp hold of them blindly and use them for social gain, ignorant of their lack of scientific evidence.

Sex differences are a typical case of this. Various branches of biology (behavioral genetics, sociobiology, molecular genetics, and neuroscience) and psychology (evolutionary psychology) have entered the great arena together to fight for the right to proclaim the "true causes" of male and female behavior. In doing so they are revisiting the sites of their scientific predecessors. The starting points for these debates were not overwhelmingly impressive. One can barely believe that less than a hundred years ago women and men were seriously thought to be at different stages of evolution, both within the same species! The development of the brain and behavior in women was seen as being "arrested" at a stage closer to the ape than to that of men. Allocation to a lower position in society, along with anyone with darker skin, seemed justified (discussed by Gould 1981). Women were said to lack the area of the brain in which the intellect resides and so were denied access to higher education and the professions. Later we will discuss these old ideas in more detail; mentioned here they serve only to exemplify that biological concepts of the body and brain have long been used as arguments to shore up support for oppression along sexual and racial lines. We may regard these as extreme positions but we need to remind ourselves that they are not far behind us in the passage of history. We live with their heritage and witness similar ideas resurfacing today, although in different forms and with a new form of rhetoric. Methodologically, some current arguments may appear more sophisticated but the sociopolitical framework on which they are hung is the same, even though it has a new coat of paint. In essence, the thinking is little changed from that of the past.

Some sociobiologists and evolutionary psychologists attempt to say that they are not against equality for men and women and do not support racism, but it is difficult, or impossible, to reconcile this with their hypotheses (e.g., see Cartwright 2001). Other sociobiologists tackle feminism, at least as they see it, and theirs is usually a limited view, and they pronounce that it would be better to realize "how our naturally selected brain acts in the service of our genes, then perhaps they would be less inclined to endure the consequences of natural selection, a blind process that cares not a whit about human beings or anything else" (Alcock 2001, p. 215).

As a basic premise let us restate that the two sexes are biologically different in their reproductive functions. The latter are easy to describe, even if the physiological processes involved are complex, and can be discussed impartially without any need to draw on complex theories pertaining to the social world. But we can no longer maintain this impartiality once the individual is born and labeled "female" or "male" in human society. From here on, biology and experience interact and attitudes begin to bear on any observation. "Female" and "male" become at once categorical divisions regardless of similarities that may stem from a shared biology. The terms are culturally overloaded with centuries of meanings and assumptions, with sanctions, obligations, and a multitude of expectations. Expectations and values may change over time but they are still tied specifically to "femaleness" and "maleness." This dichotomous cultural construct is still so firmly in place that it guides society and the science conducted within that society. The question we will address in later chapters is how much basis the femaleness/maleness dichotomy has in biological terms.

Today, we have an excellent opportunity to discuss sex differences in the context of genes and beliefs, biology and society, politics and science, and to examine how to make sense of sex differences in the maze of these many crossroads. And we can do so without resurrecting the old nature/nurture and body/mind dichotomies, despite, or because of, their extremely long intellectual histories. Polarized opinions, as were invariably produced, resulted in countless deadlocks. It is not simply a matter of finding a synthesis of the extreme positions in order to appease the opposing parties, or of finding shades of gray. Such positions ultimately reconfirm, and do not challenge, the two poles—nature on one hand and nurture on the other. In fact, the entire nature/nurture debate will have to be discarded altogether if we are to break out of the deadlock, and move to relational and dialectical models that can accommodate new biological insights without leading directly to scientific theories explaining the social world.

We have chosen to focus our discussion on sex differences not only for the topic's own interest and relevance to our lives today, but also because it serves as a model for challenging the evolutionary psychologist's claims for genetic determinism of another set of problematic assertions pertaining to racial differences (see Kaplan and Rogers 1994) and so-called "anti-social" behavior. We intend to counter scientific claims with scientific arguments and will make a point of drawing out the flaws in the broader way of thinking about sex and gender. Yet even this goal on its own might be unrewardingly self-limiting from the outset. We all know that we are surrounded today by contemporaries (and seemingly a growing number of them) who are convinced by the value of reductionist thinking and such reductionism in science, unfortunately, does not stay within the small select scientific community. It trickles out into the broader public, is often taken at its face value, and can do a good deal of harm. Since the time of Francis Bacon we have believed that knowledge is power and we have invested a great deal in specific kinds of knowledge, largely technological and scientific. Science can bestow, and has bestowed, military, economic, or political power. However, Bacon's dictum is based on the assumption that the knowledge we acquire through science is true, or, more modestly, that we can turn a dream into a workable machine, mechanism, system, or form.

Throughout the history of science, many of the scientific claims have been found to be incorrect or untrue. Hypotheses get falsified and rejected, but so keen are we, so hungry has the public been made to feel for new "findings" that these findings often get reported far too soon, and their falsification later is then no longer of interest. The gay gene is one such highly visible debate. We do not have a shred of evidence for its existence today, yet in the media hype of the mid-1990s the "gay gene" was established firmly in the minds of a broader public. Such false "implants" of alleged knowledge *are* dangerous because they have been allowed to slip into discourse, thinking, planning, and culture in a way that normalizes their existence. Biological explanations of difference, based on spurious claims, are equally dangerous because of their far-reaching consequences.

Quite often, very broad theories are being proposed and these, too, get hailed as innovative and progressive, even if, at the core, they may not be. Support and enthusiasm for such new theories also do not happen in a vacuum and the implications of some findings may hit home very swiftly in the social arena. Sex and gender issues are among the most immediately inflammatory items on the list of general interest.

Reductionism, in a way, has probably contributed to the greatest advances in technology. We are now a technologically constructed world and tend to forget that the constructs are also ideologically motivated (see also Marks 2002). More importantly, in this new positioning, we are at great risk of losing values of utmost importance. Technologically constructed thinking is rarely lateral or critical across the social spectrum, let alone broadly humanitarian. We stand to lose ethical values, most of all, because technology and reductionist thought is amoral, and much of that kind of thought has now also made its triumphant appearance in many academic disciplines. The ethical dimension of science and technology is brushed aside as if it were an unnecessary adjunct, a sentimental issue of little concern. Discovery is said to be progressive (even if the discovery turns out to be false) and "progressiveness" is such a cherished value that it can, but must not, annihilate the question: Can what we are doing become morally reprehensible? These concerns are being voiced now and, thankfully, occasionally even by prestigious organizations (e.g., Nuffield Council on Bioethics 2002).

Evolutionary psychology and molecular genetics, as academic fields, have particularly captured the research dollars and the minds of many, despite their many flawed and grandiose assumptions. However, as George Guille-Escuret and Ann La Vigne said in 1997, a pursuit of drawing up a list of their errors alone would barely make a sustainable objective for a book. Although we will engage particularly with their flaws of argument in great detail throughout the book, this is not our single objective. Our objective is to engage with the broader world and its thinking and how this pertains to our values and future. Evolutionary psychologists and a host of related academics do not alone bear the brunt of the climate of opinion that has formed around us. It is not just that we are flooded with reductionist claims because individuals have initiated it, but that the current climate of opinion is *favorable* to reductionist thinking and has encouraged it. We are trying to examine in this book how much of what is claimed about sex differences, homosexuality, and a host of other "conditions" is actually verifiable, what other explanations could be given, where we should remain skeptical (the invasion of ideology), and also ask to what possible ends some findings are being sought.

BELIEF IN GENETIC EXPLANATIONS FOR HUMAN BEHAVIOR

Current research in molecular genetics has led to exciting developments, among them the process of mapping the human genome. But what will be the outcome of this achievement when complete? The possibilities of discovering the causation of certain diseases and of tracing human ancestry and origins are, it seems, now at our fingertips and offer exciting vistas for future research. Yet gene-mapping technology has little to offer when it comes to linking behavioral characteristics to particular genes. And why *should* it have anything to offer in this regard? Knowledge of the separate fields of behavior and genetics already indicates that we should not expect to find a single gene, or even a sequence of genes, that is responsible for any specific pattern of human behavior. This is the case for the complex sets of behavior that we have labeled as gender specific, or as homosexuality and so on. Complex behavior must involve not only the expression of many and various genes but it is also shaped by experience and learning. To understand how these factors interact in an ever-changing way requires study of the processes of development, rather than simply trying to tie the expression of a gene, or set of genes, to some sort of fixed end point said to typify the behavior in question.

Evolutionary psychologists postulate that there are genes that *cause* particular patterns of behavior and molecular geneticists attempt to find those genes. Although it might be recognized that the hypothesized genes are expressed only in certain contexts or circumstances, the belief is that genes ultimately play the determining role. DNA (deoxyribonucleic acid), which makes up the genes, is seen to provide our basic building blocks for life, including our psychological and social life. In a leap of faith, a causal connection from genes to behavior is assumed, although the evidence obtained tells us nothing about the direction of influence in any association between genes and behavior. The unidirectional approach of genes-to-behavior is a beguiling way to look at a living organism but that does not make it correct. Behavior itself can affect the expression of genes. As research on animals has shown, when we behave in particular ways, specific regions of the brain are active and specific genes are expressed in ways that affect how the nerve cells function; in fact, the turning on of genes is an essential step in the many cellular processes involved in laying down memories and in altering the growth and organization of nerve cells. It may be interesting to find certain genes being expressed in a group of individuals who behave in a particular way, but this, by itself, gives us no information about whether A (a gene) causes B (a behavior pattern). Both genetic and nongenetic (environmental) factors contribute to the manifestation of any physical or behavioral charactaristic. In fact, the environmental factors are diverse, changing, and may even interact with each other. The genes are only a part of the entire process.

The sociobiologist Edward Wilson states in his book, *Consilience: The Unity of Knowledge* (1998, p. 182), "causal events ripple out from the genes to the cells to tissues and thence to brain and behavior." This statement implies that the genes are the harbingers of all that will emerge in the individual, from his or her physical type to complex behavior. Richard Lewontin (1991) refers to such assumptions as "the doctrine of DNA" which also pertains to the way in which the idea is put forward in the media and incorporated into public consciousness. To give Wilson his due, he has now moved on to including epigenetic rules that channel or constrain the path of development in particular directions and allow genes and culture to co-evolve (Wilson 1998), although his primary emphasis remains on the genes. Most evolutionary psychologists fail to consider seriously environmental (often cultural) influences and may, only when pressed, pay some lip service to their importance. The relationship between gene expression and behavior is, as we will discuss later, much more complex than an assumed unitary causation from genes to behavior.

One of the forces responsible for the new moves to pin complex human behavior to genetic causes is the pressure of society on science.

Science is not a search for truth operating in the solitude of the laboratory. Scientists are part of the society in which they conduct their research and they share its biases and prejudices (Rose and Rose 1969). And there is another problem: public hopes, expectations, and demands (and the structure and means of obtaining research funds) have edged the hands of the media to the laboratory benches where they are ready to catch tidbits before the facts are assembled. Moreover, even when findings are ready for scrutiny, it is the simple and dramatic line that fires the imagination. Complexity is not a message that captures the attention of the media, the public or, indeed, many scientists.

Powerful as the simple message of genes-to-behavior may be, it tells a tale of deceit. Influences from the environment are usually seen as minimal, or, if they are taken into account, they are seen to be constrained by the genetic code. The latter means that genes are considered to determine even the environment in which a person "chooses" to live. In fact, Edward Wilson believes that genes bias the "evolution of culture" and that "culture helps to select the mutating and recombining genes that underlie human nature" (Wilson 1998, p. 182). Gone are the continuous and changing interactions of an organism with its environment throughout its development and the rest of its life, as if they were merely a superfluous complication of the plot. Yet these are the essential processes of living organisms. Steven Rose describes it succinctly as "organism and environment interpenetrate [and] organisms actively select environments just as environments select organisms" (Rose 1997, p. 307).

The complexities of such interactions are all too readily overlooked because there is a tradition, especially in the lay media, to foster little vignettes of supposed truths. This preference comes from a long positivistic intellectual tradition and from human impatience (if not hunger) for simple and polarized explanations. There seems to be an overwhelming need in the modern era to cling to simplified explanations or accept unproven (even unlikely) causes in order to feel secure, or to feel that one's actions are justified. Today, it is as if the human condition may somehow be explained or "solved" in a test tube. Molecular genetics may lead to cures of some diseases, and so some crippling conditions may be removed from the long list of human suffering. Every such success is worth celebrating, but there is a very long road from disease, as conceived within the medical model, to the social and behavioral world of everyday life. To collapse explanations of the latter into single causal steps based on genes is to underestimate the enormous distance between finding the causes, and hence cures, for physical problems and identifying causes of psychological, social, political, and cultural issues, some of which have been designated as problems or as antisocial. Yet people often speak of genes as if they hold the program for an entire individual. This is dangerous because it implies that anything about this individual can ultimately be manipulated, just as one manipulates health and illness. So one may (one day, some hope) genetically manipulate aggression, intelligence, schizophrenia, alcoholism, homosexuality, and even gender differences (see Hamer and Copeland 1998).

Taking schizophrenia as an example here, earlier research claimed to have found the gene, or genes, apparently causing schizophrenia but this claim has met with mixed success in attempts to repeat it: Schizophrenia is a condition that cripples the lives of vast numbers of people; it would be exciting if one single gene, or locus of genes, could be held responsible and if its expression could be manipulated. The hope that there might be a cure based in identifying postulated genes is so strong that the search continues, as demonstrated by a recent scientific paper on a genome-wide scan conducted in a nationwide study sample of schizophrenic families in Finland. The scientists reported the discovery of two loci (sequences of genes) on the chromosomes that, they claim, make individuals susceptible to schizophrenia (Paunio et al. 2001). At the same time, a massive and costly effort funded by no less than six countries has reported its failure to find a gene sequence for schizophrenia (Levinson et al. 2002). Other recent papers claim to have found evidence for the inheritance of some sort of general intelligence, possibly manifested as speed of processing of perceptual information and ability to use memory (Hansell et al. 2001, Luciano et al. 2001, Posthuma et al. 2001). It remains to be seen whether these reports will stand up to further scientific scrutiny; such scrutiny is required, given the difficulty in controlling for the many factors that could bias such results. The question is why so much energy has gone into finding genetic causes for an extraordinarily wide range of human attributes, ranging from illnesses to psychological issues and even to lifestyle, at the expense of alternative approaches.

Research seeking genetic causes of various patterns of human behavior receives priority funding by governments and private bodies alike. It is as if a whole range of complex behaviors might soon be read off from the genes, like a recipe read from a book. In fact, the announcement in June 2000 of the rough draft or first assembly of the map of the human genome was couched in terms that described it not merely as a recipe book but as a "glimpse of an instruction book previously known only to God" (Francis Collins, quoted in *The Australian* 2000).

Many molecular geneticists, as well as politicians, journalists, and lay people, hold fervently to a belief that mapping the human genome will solve the problems of the human world. They do so to near biblical proportions, believing that, in the mapped human genome, they have found "The Book of Life" or "The Code of Codes," as if they held the ultimate solution to the riddle of life (Nelkin 2000). This new reverence is an ardent, and thus partly irrational, perspective, derived perhaps from very deep emotional needs in postmodern societies. Just as postmodern schools of thought have emphasized uncertainties, ambiguities, and fluidities, and have proceeded to debunk and unsettle (deconstruct) our modes of thinking, some aspects of scientific and popular culture have moved in the other direction, proclaiming that solutions and certainties are around the corner in areas where they might not or need not exist. While one may sympathize with the desire for certainty, it is manifestly a hope that has turned our eyes away from those dimensions of the individual that depend on experience, insight, contemplation, and the complex and dialectic interaction with the environment.

This particular way of thinking believes that all of the recipes for all complex behaviors are already, at birth, in the individual's "Book of Life," waiting, as it were, in readiness to be expressed at a certain stage of life or in a certain range of environmental conditions. Hence individuals have predispositions to behave in particular ways or, as the above-mentioned Finnish study of schizophrenia puts it more negatively, they have "susceptibilities."

Believing thus, on a broad range of perceived susceptibilities, opens the way for a number of social actions. One is to test fetuses and abort them if they show certain signs of perceived potential for negative traits. Another is to manipulate the DNA of fetuses to eliminate or minimize risks and hence to eventually make certain susceptibilities disappear from the gene pool. It then also allows the screening of individuals to look for "genetic susceptibilities." Once they are pinpointed, the optimists claim that we will use the information to create environments to ensure that individuals with genetic susceptibilities for undesirable behavior will not express that behavior. Others see this information being used to exclude "susceptible" individuals from certain professions, insurance programs, lifestyles, and so on.

Let us take a hypothetical example. A pregnant woman is told that the embryo has a "potential" for homosexuality. Would she choose to abort, or be pressured to abort, or be made to feel that she is irresponsible if she carried the pregnancy to full term? Even if she or both parents decide to allow this pregnancy to proceed, how would their attitude to the child change even before its birth, and how would they handle the raising of this child? In what environment would you raise a child carrying the "genetic susceptibility" for homosexuality, assuming that there is such a gene sequence (which we discuss critically in Chapter 9) and that being homosexual is considered undesirable (which we do not accept)? We can only imagine that such a proposed environment might mean locking the child away from a broad range of life's experiences, giving the child a very different upbringing than he or she might have received otherwise. We can also spin the argument further by assuming that screening for negatives will not stop at one proposal (e.g., homosexuality) but proceed to broaden the definition of what constitutes an undesirable genetic composition. Attempts have already been made to pinpoint a genetic basis for criminality, so-called asocial behavior (including not paying one's bills or being homeless), and violence. Once undesirable behaviors have been singled out for genetic examinations (even on spurious evidence), we might then routinely insist on preventing whole classes of people from breeding, by routinely sterilizing them, or forcing them to have abortions, with no consideration for the social conditions that might have produced such behaviors in the first place, as contrasted to supposed individual genetic susceptibilities. The latter is, of course, highly attractive to many Western governments. It assumes that by genetic manipulation one can control social ills genetically and avoid frictions of beliefs and norms.

It might seem far-fetched to think of all members of the society being screened for claimed genetic susceptibilities, and so carrying around their own recipe books for traits they *might* express if we do not watch out, but some scientists and policy makers believe that the potential is just around the corner.

The Human Genome Project has finished mapping the locations on the chromosomes of all of the close to thirty thousand genes that make up the human genome. Many of those genes code for certain physical features and particular aspects of a person's physiology. There are clear cases, known for many years, in which single genes influence features, such as eye color and hair color. These features depend on a particular gene making a single protein. Some genes code for proteins that are enzymes and some of those enzymes are important to the functioning of the brain. If a person has a genetic defect that prevents the body from making one of these enzymes, brain function can be affected in many ways. For example, phenylketonuria is a genetic condition in which the enzyme that converts the amino acid, phenylalanine, to tyrosine cannot be made and a broad condition described by the collective term "mental retardation" results, unless tyrosine is supplied in the diet from birth on. It is possible to find out where on its chromosome the gene responsible for making this enzyme is located. There are other examples of specific genetic defects that affect behavior in very general ways. But it is not so easy, or even possible, to find out whether a single gene affects behavior more specifically.

In fact, despite the hopes and claims of molecular geneticists, it is most unlikely that any particular behavior depends solely or largely on the action of a single gene, or even a string of genes and, therefore, it is far from likely that a gene map of units of human behavior will be discovered. While the expression of genes is a part of the process, their input is usually so malleable, depending on interactions with the environment, that it would be impossible to make any direct association of the behavior with a set of genes, let alone a single gene. In other words, there is no chapter in the "Book of Life" with a recipe for understanding the causes of complex human behavior.

Of course, it is an important aspect of discovery to dare to dream, to cast the net wider and to think beyond what exists. This is the essence of science and discovery. Science fiction writers have often proposed models and ideas long before the remotest chance existed of putting their fantasies into practice. However, much of that dreaming today is turned into rigidly held beliefs and categorical statements imparting a false reality. Beliefs are stated as facts or entities very soon to be proved. Some scientists have even declared that the nature/nurture debate (that is, experience as opposed to genes) is now over because genes have "won" since genes have, supposedly, been shown to control human behavior (e.g., Horgan 1993, p. 93, citing the editor of the journal *Science*).

As a consequence of this belief, they attempt to convince us—and funding bodies—that we are on the verge of eliminating many problems of society. "Quality citizenship" and quality "world citizenship" are new catchwords. Behind them hides a new eugenics and one that is, once again, based on pseudoscientific justifications. For instance, Raymond Cattell (1987) of the University of Illinois wrote, "Eugenics favors preventing births of those who would inevitably be miserable and incapable of living a normal happy life" (cited in *SPLC* 1990, p. 30). It is now a question of what society would call miserable, normal, and happy. In a similar vein, Edward M. Miller (1997) of the University of New Orleans argued that we should be moving toward a world in which births are actively prevented "that are most likely to result in what that society regards as low quality citizens," and that we have a choice of "having the population restrained by misery, and having it restrained by conscious restrictions of births" (cited in *SPLC* 1990, p. 30). Savulescu argued in *Bioethics* (2001) for a eugenic selection of embryos, to eliminate not just heritable diseases but possible shortcomings such as low intelligence. He was careful not to mention homosexuality but the principle he defends as "procreative beneficence" (selecting the best children) fulfills the requirements of former principles of eugenics, which have been shown to be no more than blueprints for sinister control mechanisms of societies, even to the extreme of genocide and for this we have irrefutable evidence (Chorover 1979, Darrow 1926, Gosney and Popenoe 1929, Kaplan 1994, 2000b).

NEW TECHNOLOGIES

Prenatal diagnosis is now used widely to detect spina bifida and Down's syndrome, followed by termination of such pregnancies, but use of such a technique to eliminate nonclinical characteristics would not be acceptable on moral grounds, even if it were possible. The same applies to the use of the new technique of preimplantation genetic diagnosis, which screens embryos obtained for *in vitro* fertilization (IVF) before they are implanted. Whether one chooses to terminate a pregnancy or prevent one from occurring, of course, raises different ethical issues. It is not our interest here to debate such issues on abortion, but rather to point out that *selection* of embryos with specific behavioral characteristics that are merely a matter of preferences biased by racism, sexism, homophobia, and so on, would indicate very dangerous trends in a society.

Additional to the new techniques in molecular genetics, other advances now allow us to probe the living brain and reveal how it functions. Brain scans provide computer images showing the blood flow and the activity of nerve cells in different regions. Hence, we can see what parts of the brain are active when a person performs a particular task or thinks about certain things.

The new imaging techniques have increased our understanding of brain function in a number of ways and have been beneficial in medical therapy. For example, imaging of the activity of the brain of a person who has had a stroke can provide valuable information about the damage that has occurred because the image shows where the blood supply is cut off or where the nerve cells are inactive. Carrying out the same procedure on a person suffering from epilepsy can reveal where the seizure starts and whether it is localized or spreads across the brain. In this case, we are seeking "red spots" of high activity in one part of the brain and seeing how the activity spreads to other regions. There are many other direct medical applications of the new procedures, used in the hope of diagnosing debilitating conditions and finding new therapies.

Other techniques allow us to make fine measurements of the sizes of structures deep within the brain and to measure the levels of the sex hormones circulating in the bloodstream and in the brain itself. We can also use a technique called "finger printing" to determine what genes are expressed in the brain, as well as in other parts of the body. This very new technique determines which genes are being expressed at the time the sample of tissue is collected for assay; this is important because far from all of the genes are actually expressed at any one time. Exactly which ones are expressed will depend on which ones are available to be expressed, changes that take place during development as well as other stages of the life cycle, and the influence of experience in the individual's past and around the time the sample of tissue was collected.

Not surprisingly, these new technologies for examining the brain are being used to study sex differences in brain activity, gene expression, structure, and neurotransmission. But unlike the results of research on abnormalities in brain function, the results of research on sex differences produce no obvious positive outcome. Most of the research in this area falls into the category of research for new knowledge. Searching for new knowledge is, of course, highly desirable and should not be constrained by any need for a direct application. But in the case of researching sex differences, we find that the experimentation and interpretation of the results is often distorted by social attitudes and used falsely to justify inequality. For example, finding that women and men have, on average, different patterns of activity in the brain when they perform mathematical tasks is usually seen as a justification for male superiority in mathematics and for employment of more men in professions depending on this ability.

It is a regrettable fact that perceived biological difference is usually used to divide one group of people from another. In addition to the study of sex differences, the now substantial amount of research comparing the fine structural aspects of the brains of heterosexuals and homosexuals, or heterosexuals and transsexuals (or transgenders), is not simply a matter of scientific interest without social implications.

NOT JUST GENES

To understand what causes sex differences we must attempt to bridge the gap between biology and psychology, and even between these two disciplines and sociology. Research in this area covers, at one level, the more tangible elements of the brain, including its structure, chemistry, nerve cell interactions, and the cellular expressions of the genes. It also includes the far less tangible manifestation of human behavior described and analyzed at different levels—in social terms and also in terms of the individual's psychology. To this we must add the complex and dynamic interdependence of all these levels.

To say this in another way, there is a hierarchy of orders of analysis that applies to such research, each level corresponding to the traditional disciplines: physics, chemistry, physiology, psychology, and sociology. This order is one of increasing complexity. We might, for example, discuss sex (or gender) differences in behavior at the sociological level as an issue for society, or we may choose to discuss the same topic at the psychological level and so deal with comparison of the behavior of individuals. Or we might look at the physiological aspects of sex differences in behavior and even consider them at the molecular (for example, hormones and genes) or sub-molecular level. As Steven Rose (1984) has pointed out, all of these levels of analysis can coexist in parallel, but there would be something incorrect if the explanation at one level contradicted the explanation at another level.

At the higher levels of complexity, behavior involves the interaction of the whole individual with the environment and with other individuals. The level at which genes and hormones operate is on a lower level of the hierarchy. Although there is no reason why we should not discuss behavior in association with, for example, genes or hormones, we must recognize that these explanations are not complete in themselves. In other words, trying to explain behavior entirely in terms of hormones or genes requires oversimplification and so distorts the picture that we lose sight of many of the important factors that also influence this behavior. By doing this, we have reduced the explanation of behavior to a lower level of analysis. This form of thinking is referred to as reductionism.

To exemplify what we mean by reductionism we would like to use an example that pertains to women. Depression in women is often said to be caused by an imbalance of estrogen or progesterone and variously termed premenstrual depression (or menstrual or postmenstrual depression), postpartum depression, or postmenopausal depression. In most women who suffer from this condition, psychosocial factors are just as important as the hormonal ones, although they are often ignored. For example, it has been shown that attitudes to menstruation stemming from what a girl learns about it before puberty influence the occurrence of premenstrual depression (Rogers 1979). Ignoring such factors means that the complexities of the mental state that we call depression are reduced to the action of one or more hormones. They may be further reduced to the action of certain molecules within the brain: for example, to being caused by an imbalance of one of the neurotransmitters, either noradrenaline or serotonin, by which nerve cells communicate with each other. The fact that the symptoms of some forms of depression can be reduced by treatment with antidepressant drugs (the tricyclic antidepressants and the monoamine oxidase inhibitors), which act by altering the activity of noradrenaline and serotonin in the brain, is often used as evidence that depression is caused by an imbalance of these chemicals in the brain. But the drugs may be having their effects by acting in ways that are unrelated to the actual cause of the depression. A certain event, perhaps long into the individual's past, may have triggered or had a role in the onset of depression. Added to this, states of mood that are labeled as depression in one society are labeled as another mental condition in another society. In other societies, the same mental states may even be assigned as a religious inspiration or curse, rather than being seen as one of the collection of mental conditions. It follows, then, that it is simplistic to describe depression only in terms of noradrenaline and serotonin, or in terms of hormone levels, although that is often done in medical textbooks and many research papers in the field of pharmacology and, at this level, is also an inadequate description.

Emphasis on the physical aspects of a disease, trauma, or other clinical condition is typical of Western medicine, but the treatments employed may not be concerned with cause and effect. To give an example, a physician who is pumping out the stomach of a female patient in order to eliminate or minimize the effects of an overdose of potentially lethal pills usually has no time to sit with the patient after the crisis is over and discuss the causes of her attempted suicide. Unless the patient is well insured and in a well-provisioned hospital, there may be no one to speak to her and, when she is well enough to go home, she may well return to the same situation that caused her to decide to end her life in the first place. In other words, not all treatments deal with causes and not all causes are physical, even if the manifestations may be. There is little point in talking about hormonal imbalance when a woman has just been left by her husband, has received notification of eviction, has no income and has no one for support. In fact, these stress factors may *cause* a hormonal imbalance, not vice versa. This seemingly simple example illustrates that a complete understanding of any aspect of behavior and its development depends on trying to establish all the factors that have contributed to the continually changing and interacting system.

FREEDOM FROM GENES?

Belief in the possibility of human freedom of action beyond a mere fatalistic playing out of a genetic program is worthy of consideration at a time when the predominant interest in genes has spawned so many self-delimiting theories of the human species.

One of the first responses by Voltaire after he had visited Newton was to drop the subject of free will as unproductive for scientific inquiry. Henceforth, he decided that it was intellectually more profitable to look at how things work (empirical observation) rather than at what they are (inferring an identity) (Libby 1935, Wade 1941). The Anglification of French thought in the eighteenth century led to substantial advances, and science was cross-fertilized across the English Channel by the collaboration of these two giants of ideas—Newton in England and Voltaire in France (Gooch 1956). However, although the idea of free will had little presence beyond Hegel and the nineteenth century, it did not die entirely.

In ancient Greek philosophy, specifically for Socrates, "will" denoted a degree of freedom that an individual obtained by intellectual insight. The freedom lay in making rational decisions based on having at hand the right knowledge and experience. The wise person, according to Socrates, is free presumably because he or she has a wide range of options to act and to choose. The later scholastic tradition (e.g., St. Augustine, 354–430 AD) regards will and intellect as the two rational faculties of the human soul, and scholastics draws a distinction between animals and humans on the grounds of intellect and will, arguing that animals must obey the laws of nature, whereas human beings are free insofar as they can direct their actions according to their own choice. Of course, animals also make choices and some, as we now know, do possess higher cognition, but here the point is whether "choice" is a matter of free will or genetically determined.

The expression of genes is, of course, essential to the physical existence of the brain and thence the mind. Whether we see that mind as passive, as did the Gestalt psychologists, or as active and dependent on conscious and unconscious agencies, as did Freud, this in itself does not dispute the concept that our mind and, by extrapolation, our actions are shaped and predetermined genetically. However, in the social sciences and in law, there is the assumption that the human mind contains the possibilities of reflective intelligence and moral decisions and in such reflection it becomes a social process (Mead 1934), aided by knowledge and experience. These processes are more than the sum total of our genes, just as personality is more than our flesh and bones.

C. H. Cooley's views were not very far removed from those of Socrates when he argued in his classic text of 1922: "Intelligence, in its fullest sense, is wisdom, and wisdom draws upon every resource of the mind." That is, for those who believe in freedom to act and in the possibilities of being an active participant in the world and in history, individuals themselves can direct voluntary actions and courses of thought.

Deterministic theories (as we shall discuss in the next chapter), on the other hand, leave no freedom to act in the broader sense. Claiming that homelessness is the result of a particular gene or that someone cannot help being aggressive because he happens to be male or has a particular gene denies the human ability to draw on past experience, knowledge, and other intangibles such as love or self-confidence, for example. It denies our having choices that are not already preprogrammed in the genes. Even more important, a specific context of events or circumstances may make us do things that we might otherwise not do. Human animals and all living organisms have the capacity to modify their behavior and such modifications include actions. Some will do so better than others but it would be difficult to argue that the choices made are *directly* explicable by our genes.

Sex differences in the genitalia or some of the so-called secondary sexual characteristics *are* largely genetically and hormonally determined, and this physical division alone alters the social conditions in which the individual must live. Yet even within genetically predetermined conditions, there are almost infinite choices as to how we deal with that basic, physical condition. The roles imposed on girls and boys (and women and men), deciding what is acceptable behavior into the very core of our most intimate lives, could not be genetically predetermined—otherwise how could women have changed their roles so rapidly in response to the women's movements? Some may respond by saying that, although there is some leeway for changing sex roles, the genes set an upper limit beyond which no further change is possible. This sounds like a reasonable counterposition. However, it confuses the biological limits (such as the ability or inability to bear a child) with sex *roles*. The latter is a social construct, as the term "role" clearly implies. Some social roles are linked to reproductive physiology, but these specific differences have been exaggerated to claim that they underpin sex role differences in vastly diverse social contexts even though these have nothing to do with our genes expressing themselves. As we will show in later chapters, some evidence indicates that the flexibility of the brain and sex differences may be far greater than some people expect or are prepared to admit.

In discussing this topic, we are largely dealing with belief structures rather than scientific evidence. Biologists are often far too eager to intercept social processes bending beliefs into scientific "fact." Biology, as Hilary and Steven Rose rightly said, is a historical science shaped by social values (Rose and Rose 1969). In this self-appointed role, biology has played many sinister roles in the past. It has tried to persuade us, often in concert with anthropology, that some human groups are inferior to others. It has tried to prove over and over again that women are intellectually, and even morally, inferior to men. Most biologists and psychologists are not trained in sociology but some of them try their hands at grand social theorizing (e.g., Dawkins 1976, Rushton 1994). These authors present roughly hewn, untrained attempts at sociology, yet believe that their expertise in biology provides suitable qualifications for pronouncements of vast political and social consequence. Such expertise, when translated into broad social issues, is often nothing more than pseudoscience and very poor social science. This is quite different from arguing, as some others might, that scientific inquiry into those fields should stop for fear of how it may be applied in the social, political, or even psychological arenas.

On the contrary, we argue in this book that we know far too little about the social and psychological implications of sex differences to make grand assumptions. The more carefully we tread, and the more focused and modest we are in staking out scientifically well controlled (limited) questions, the more we may advance our knowledge in how extensively we are free agents (philosophically speaking), determining our fate by a dynamic, even dialectic, process throughout life. In this way we include development, experience, learning, and a multitude of significant processes that lose their importance in constructs of genetic determinism.

2

Essentialisms, Determinisms

Essentialism and determinism are separate philosophical positions that have led to some significant advances in science and ideas in general, but these positions have often been turned into dogma and ideology. They have been used inappropriately (illogically) to support ideological positions for which there is no good scientific evidence and they have added to the oppression of certain groups in society. We will elucidate why essentialism and determinism have rigidified debates on gender, genes, and other social markers such as ethnicity, leading, in their worst application, to sexual and racial prejudice.

Looking at the history of modern science, we can see that the essentialist intellectual position has remained a very important influence on the way we view science today. Essentialism argues that there are categories of objects and genres that have essential characteristics, notwithstanding individual variation, and that these essential characteristics define the objects and genres to an extent that they reveal truth. Abstractly, it is also possible to have universals, entities that might be revealed briefly to confirm their universality. John Locke, for instance, sketched a tentative thesis of science on the basis of essentialism. He argued that objects have essences and that these can explain their observable properties. The ultimate goal of scientific investigation had to be based on uncovering the real essence (truth) about the object. For instance, a drop of water may be very different in appearance from an ocean, but in essence, both share the same chemical properties. In discovering this composition, we can then argue about its appearance, but the characteristics of water remain the same. In searching for principles of this kind, this way of thinking opened the doors for systematic investigation of substances and their properties. Substances no longer had to be imbued with a will of their own. By discovering not so much their essential nature but their observable properties it became possible to predict what the substance would do under certain circumstances (e.g., water would boil or freeze at expected temperatures).

Essentialism, as a philosophical position of the seventeenth and eighteenth centuries, was thus a seminal position for the advancement of science as we know it today. However, when the same principles of essentialism have been applied to the social world, the claims have usually been mere assumptions and not truths; they have worked only by allegory or as metaphors. They have often had a demagogic intention. The simplest form of a demagogic application of essentialism is mentioned in the Bible (Jeremiah) when the often-repeated question is asked: "Can the leopard change his [*sic*] spots?" (Stanton 1960). Presumably, the answer one should give to this question is "no" because we all know that the *individual* leopard cannot change its spots. By saying "no," we support an essentialist position. The essentialist denies flexibility in the individual and its ability to change in response to the environment but this may not prevent an essentialist from accepting that species can change through evolution. The theory of evolution maintains that, whereas every individual within a species has essential characteristics typical of the species, genetic selection by particular environmental conditions can favor those individuals with mutated genes (changed essences) so that in time, and occasionally even within one or a few generations, the species as a whole changes to express a new essence, eventually to become a new species (e.g., no longer a leopard species). This perhaps seemingly convoluted discussion gives us the perspective to see whether the sociobiologist or evolutionary psychologist of today is an essentialist. In our opinion, the positioning of both these groups is consistent with essentialist thinking. They both consider the genes to be the essence of the individual.

Determinism differs from essentialism, but it overlaps it to some extent. Determinism takes essentialism several steps further by actually looking at the natural world in a different dimension. If essentialism is the structural way of explaining objects, determinism is its dynamic component. Determinism is a theory (by now a whole set of theories) or doctrine(s) that everything in the world and in nature is subject to laws of causality. Everything is as it is and will be by virtue of some laws and forces that make it so. At the simple level, using the example of water again, the boiling of the water is determined at a certain temperature because molecules begin to move further apart at that temperature. Hence, the water cannot do anything other than boil if we expose it to temperatures of around 100° C. Determinism would simply argue that any event that has happened had to happen and any event that has not happened could not have happened. By arguing that there is a reason for everything, determinism has engendered the search for causality as a positive dimension of science. But there are exceptions: even quantum physics does not subscribe to causal determinism. Rather than state that the ultimate laws of nature are causally determined, quantum physicists speak about statistical probabilities of occurrence at the subatomic level. It is thus possible to describe cause and effect in other ways. As a particularly poor application of deterministic thinking, the search for causality has led to a reduction of possible causes (or interplay of various elements) to one single alleged cause: hence the term *reductionism*. Invariably, this way of arguing is guilty of reducing complex phenomena to a singular cause, thus resulting in distorted or even misleading concepts, ideas, and statements that are then offered up as truth (Rose 1984).

When applied to the social world, deterministic thinking is invariably reductionist (oversimplified/overdetermined). As we explained in Chapter 1, the dominant way of thinking in Western traditions of biological science is to reduce explanations to their simplest form. This process of reductionist thinking involves seeing unitary causes for complex sets of events. For example, explanations for sex differences in behavior that have given either genes or hormones a primary and overriding role are reductionist accounts because they simplify (that is, reduce) complex interactions to a single causal event. They either partially or entirely ignore the importance of experience or stimulation from the environment surrounding the organism. Deterministic attitudes present false or unproven assumptions of causal relationships and thus grossly distort the complexity of the chain of events. It is then a small step to argue that the causal relationships reveal an immutable essence of the individual organism.

There is a further, more sinister, ethical dimension to determinism. If every person is causally capable of acting in only one way given a certain set of circumstances (i.e., has no choice), and the way we act as a species is the only way we can act, then we are not morally responsible for our actions. For arguments of gender this is a very important point and we will come back to it in this and later chapters. Great thinkers, like Hume, fully recognized the risks of determinist thinking and, therefore, considered that determinism should be applied only to the most general laws of nature. Clearly, Western traditions have constructed laws that depend on humans (at least of adult age and of sound mind) being considered responsible for their actions. We can be punished for defiance of the laws because it is assumed that we have a choice of acting or responding in a number of different ways, one or some of which would not be breaking the relevant law. Outside the law courts, in gender relations, assumptions are often held as if there were an essentialist "nature" distinguishing men and women and that they have separate needs based on their biology.

In human society, essentialism is usually associated with gender, race, and, to an extent, even age. In fact, woman and man are the most essentialist categories we have devised. Vocabularies and even the semantics of languages are built around this division, and there are only a few languages that express a male or a female by a third-person pronoun. Babies are sometimes called "it," as if to postpone gender identification, but it is almost impossible to speak about a third person without referring to a gender-based pronoun such as "she" or "he." This becomes noticeably problematic when the speaker does not know the person's sex. We cannot escape linguistically, and therefore not easily in thought, the idea that things are as they are and will be so forever more. People employ essentialist categories when they talk about men, when they say that something is typical for a woman, or for an eighty-year-old, a black, or an Asian. It is assumed that once one knows one eighty-year-old, one knows them all, and so on for the other labeled categories. This may sound rather obvious, but it is by no means a trivial matter when these categorized and generalized attitudes operate in social life.

Racism, antisemitism, and sexism would be unthinkable without essentialism, and most of these divisions are based on the same logical flaw. This flaw is the assumption that each category is somehow reproduced almost identically from one generation to the next and that the differences between members of the same category (such as women, Asians, blacks, and old people) are perceived to be negligible enough to warrant extreme generalization, as if a description of one were sufficient to describe all. But essentialism is more than simply an assumption that characteristics can be generalized. It is a means of drawing closure on a subject because it is presumed to be understood without requiring investigation. The judgments deriving from such positioning and posturing are thus prejudicial. We shall give some examples relevant to debates on sex differences throughout the book.

Determinism in relation to the categories male and female follows a circular causality. Allegedly, we act as we do because we are female or male and, since it is already known that a specific behavior is attributed to a male or female, this then makes it typical. For instance, according to this view, chromosomes account for the fact that an individual is female, but then femaleness is not a descriptor of the chromosomal arrangement but of every act of behavior emanating from the individual so designated. She is caring because she is female. She is anxious because she is female. To expose this as circular reasoning we need only question how we would be able to explain causally the possibility that a male is caring. We can say only that he is an atypical male, abnormal, or even deviant, but we cannot give a causal explanation based on his gender because we have already causally related such behavior with femaleness. We come across such logical inconsistencies regularly in everyday life.

Biological determinist theories, in their historical context and social ramifications, are part of a wider set of constructs that attempt to justify the oppression of races, classes, and minority groups (Lewontin et al. 1984). They are also poor science and do not foster our understanding of either individual or group behavior. Two related paradigms underscore all such thinking: reductionism and division into dichotomous categories—black versus white, male versus female, maleness versus femaleness, male sex hormone versus female sex hormone (Birke 1982). As a polarization with discontinuity, a dichotomy ignores overlaps or gray areas. Differences are seen to be more interesting than similarities and there is a tendency to see these differences as absolute. As a consequence of this thinking, the psychological literature has placed undue emphasis on sex differences in behavior at the expense of similarities.

SEARCHING FOR DIFFERENCES

The belief in essential differences between female and male has a very long history and one that would lead us too far away from debates of essentialism and determinism in modern science to be covered here. However, we need to mention that, in science, the search for differences between the sexes received a new impetus in the second half of the nineteenth century, at the very time when women in the bourgeoisie had acquired a new respectability and began to show signs of emancipation. We can trace these early ideas up to the present day; even though they are expressed in different forms and under different headings they are not necessarily different in content. It is important here to iterate that science, despite its claimed search for truth, universality, repeatability, and alleged value-free approach to its objects of study, does not exist in a vacuum and is, in fact, swayed by social values and attitudes. In nineteenth-century Europe, debates were conducted in an intellectual environment in which slavery had been abolished but imperialism was at its height, in which rapid industrialization and major advances in science were matched by a new set of formulas on how best to justify colonial expansion. Nineteenth-century science conveniently spawned theories suggesting that the brains of so-called white people were better developed than those of so-called black people, and that the male brain was better developed than that of the female. Females were said to lack the center for intellect, and blacks, too, were seen to have inferior intellect. Putting it the other way around, Charles Darwin (1871) stated that "some at least of those mental traits in which women may excel are traits characteristic of the lower races" (p. 569). He also referred to the "close connection of the Negro or Australian and the gorilla" (p. 201). To make the logical connection that he, perhaps deliberately, leaves to conjecture, females must also, in his opinion, be closer to gorillas and hence at a lower level of evolution than men.

Some of the most reputable scientists of the time concerned themselves with finding scientific evidence to support the belief in the inferiority of females and blacks. They became obsessed with measurement of cranium size and brain weight. The assumption (incorrect, as shown later) underlying this approach was that smaller brain size and weight indicate lower intelligence. Black and female brains were considered to be not only smaller and lighter in weight, but also underdeveloped. Thus the anthropologist, E. Huschke, wrote in 1854, "The Negro brain possesses a spinal cord of the type found in children and women and, beyond this, approaches the type of brain found in higher apes" (cited in Gould 1981, p. 103). Similarly, another anthropologist, McGrigor, wrote in 1869 that "the type of female skull approaches in many respects that of the infant, and still more that of the lower races" (cited in Lewontin et al. 1984, p. 143). The French craniologist, F. Pruner, typical for his time, wrote in 1866; "The Negro resembles the female in his love for children, his family and his cabin. . . . The black man is to the white man what woman is to man in general, a loving being and a being of pleasure" (Lewontin et al. 1984, p. 143). One notes here the equation of the "Negro" *he* with the white *she*. There were many such crude attempts to draw a close association between blacks, women, and apes (see Gould 1981). Females and black people were often equated to justify the domination of men over women and the domination or enslavement of one racial group by another. Thus, sexism and racism were linked aspects of biological determinist thinking (Richards 1983).

Craniology flourished in the second half of the nineteenth century, and the then famous neuroanatomist, Broca (1824–1880), founder of the Anthropological Society of Paris, gathered much data demonstrating that the male brain was on average heavier than that of the female. His co-worker G. LeBon believed that women were an inferior form of evolution, equated to children and "savages" and without the capacity to reason (see Gould 1981).

LeBon was one of the founders of social psychology. He was opposed to higher education for women, as most men of his time believed. His views were not based on some intrinsic malice (although he was as misogynist as one may get), but because of an essentialist position about the entirely different biological makeup of men and women. Remarkably, so deep was the prejudice that it went unnoticed by scientists of the time that perceiving men and women as beings on different levels of human evolution might present difficulties for reproduction.

Quite early in the era of brain size measurement scientists had run into a number of problems. For example, Broca's theory of brain size and intelligence did not match his findings that "Eskimos, . . . and several other peoples of the Mongolian type" had, on average, a larger cranial capacity than "the most civilized people of Europe" (Broca 1873, p. 38). To circumvent the problem Broca claimed that the relationship between brain size and intelligence might not hold at the upper end of the scale because some "inferior" groups have large brains, but that it did hold at the lower end. That is, small brains belong exclusively to people of low intelligence. Another problem was that the brain weights of several men notable for achievement were embarrassingly lower than the European average (e.g., Walt Whitman and Franz Josef Gall, the founder of Phrenology). Undaunted by these statistics, Broca proceeded to make excuses for these findings, claiming that those famous men with lower brain weight had died at an older age, which may have caused loss of tissue, were smaller in build, or that their brains had been poorly preserved. He did not apply this reasoning to the brain weights of the women sampled.

We know today that the majority of the brain's volume is extra-cellular space (filled with extra-cellular fluid) and that different preserving (fixing) methods of the brain cause different amounts of shrinkage. Hence we are able to see clearly with hindsight how extremely unreliable these early data on brain weight and volume were, and how prejudicial their interpretation. Moreover, the idea of brain weight as a distinguishing mark between males and females was finally extinguished when it was realized that no sex difference occurs when brain weight is expressed in relation to body weight. In 1901, Alice Leigh applied new statistical procedures to the analysis of the data and showed that there was no correlation between cranial capacity and intelligence or brain weight and intelligence. Nevertheless, it was in 1903 that the American anatomist E. A. Spitzka published a figure showing the brain size of a Bush woman as intermediate between that of a gorilla and the mathematician Gauss, describing the number of convolutions of the surface of the brain of the Bush woman as closer to that of the gorilla (see Gould 1981, Fig. 3.3). It is worth noting that the lower "white" classes were also slotted into having inferior, undeveloped brains (Gould 1981). Genetic explanations of the supposedly inferior behaviors of black people were also postulated by eugenists such as Francis Galton (1869) and Karl Pearson (1901). When brain size failed to prove beyond doubt that men and women were different (and unequal), attention shifted to structural differences in the skull or differences in subregions of the brain, such as the frontal lobes or the corpus callosum. Comparative measurement of brain regions in male and female brains is still carried out today.

Another attempt to confirm female inferiority on grounds of differences in structure and function of the brain was the so-called recapitulation theory. Recapitulation thinking tried to explain why women and blacks were allegedly childlike: the brains of women and blacks suffered from "arrested" development and were therefore allegedly not as well developed as those of white men; indeed they were said to be infantile. This led to the formulation of the hypothesis that "ontogeny recapitulates phylogeny," maintaining that the individual passes through the various stages of evolution as she/he develops (e.g., from fish to reptile and so on up the evolutionary scale as development passes from fetus to child to adult). Hence, it was thought that throughout development and growth the individual biologically repeats a series of evolutionary steps, each representing an adult ancestral form in the evolutionary order. Applied to the brain, this was supposed to show that some brains of humans were "more ancestral," meaning that they were closer to animals. Blacks, women, and the lower classes were seen as having more ancestral brains than white males and they were said to have brains more equivalent to those of white male children rather than to white male adults: that is, they were not considered to reach a fully developed state of intellect. This recapitulation view is still upheld rather widely in the Western world.

More important, recapitulation theory belonged to the category of determinist beliefs that convinced early fascist writers of the biological basis of all behavior. Rudolf Hess argued in 1934 that National Socialism was "nothing but applied biology," correctly assessing the role of science in policy formation under the Nazis. Although the Nazis did not have a coherent policy on women, they were keenly interested in sexual and reproductive behavior (Kaplan and Adams 1990, Kaplan

1994). Theirs was not so much a scientific interest as it was an irresistible ideological challenge, to create the perfect race by human intervention. These conclusions came in the wake of Darwinian theories and the discovery of the importance of genetics. The official passing of the German "Law for the Prevention of Progeny of the Genetically Unhealthy" in 1933 gave women, as childbearers, a new status within that framework. Their own genetic stock, as well as their alliances with sexual partners, became a matter of government scrutiny, usually via the medical profession, specifically doctors in the service of the Nazis as "biological soldiers" in the war against "eugenic pollution." Chorover (1979) argued provocatively that it was originally sociobiological scholarship that provided the conceptual framework by which eugenic theory was transformed into genocidal practice in Nazi Germany. Thus he established a clear link between eugenic practice by the Nazis and intellectual theorizing of sociobiology in the 1970s.

CREATING SYSTEMS OF CERTAINTIES

Theories of biology (e.g., survival of the fittest), which in themselves may have no particular moral standpoint, were increasingly translated into broad social and political thought and application. A well-known example is the way in which biological views on gene "contamination" were translated into the eugenics movement of the 1930s. From the 1960s on there has been renewed interest in biological determinist theories for explaining human behavior and the structure of human societies. These theories have been applied to race, class, and sex, but the idea that natural selection has selected for genetic differences between blacks and whites, men and women, and different social classes has not been without its critics (e.g., Lewontin et al. 1984, Rose and Rose 1976, 1978, 2000).

It seems that the more we know about biology, the greater the temptation to draw parallels with social life, as if this gives the latter scientific validity. Sociobiology came to the fore with the publication of E. O. Wilson's book *Sociobiology: A New Synthesis* (1975) and was made popular by Dawkins (1976) and Barash (1979), among others. It is a field of inquiry that has been seen to be tainted and influenced by the pseudoscientific claims from generations earlier despite its eloquence and eruditeness. It does not suggest large population control mechanisms or eugenic programs. Sociobiology does argue, however, for a direct nexus between biology and social manifestations of complex phenomena.

Similar to the molecular geneticists mentioned in Chapter 1, sociobiologists claim to explain a wide range of complicated human behavior by tracing these characteristics back to the genes and by making genes the paramount determining factor. Dawkins (1976) referred to the genes as "selfish" and as "replicator units" that control every aspect of our behavior in order to maximize their chances of replicating and so getting into the next generation. To Dawkins, human culture is merely a by-product of the replicator units' program to perpetuate themselves. Sociobiologists Trivers and Willard (1973) formulated a theory of sex bias in parental investment, stating that mothers should bias their investment in their offspring in favor of the sex that demonstrates greatest fitness; that is, more investment should be given to the sex that will produce more grandchildren. According to this theory, in species in which one male mates with several females, mothers in good physical condition should invest more in their sons than in their daughters because males with large body size will be more successful in mating. The reverse bias of maternal investment should be the case for mothers in poor condition; since their sons are not going to be large and successful males, they might as well invest in their daughters. The sexism is obvious, but what about the science? As a recent review by Brown (2001) points out, examination of the data for nonhuman primates reveals no evidence in support of the theory.

Sociobiologists are also of the firm belief that the genes determine sex differences in behavior, and so this position leaves no real potential for social change. According to Wilson (1975), genetic differences between the sexes are great enough to cause a substantial division of labor even in the most free and egalitarian of societies, precisely because the division between the sexes is rigidly determined by biological differences. He believes this to be true for all human societies.

The sociobiological theories rapidly became politically very important, as they were readily taken up by the media and incorporated into a multitude of disciplines, discussions, and attitudes. The supposed genetic basis for aggression, territoriality, and intelligence fitted well with characteristics of the stereotypic male of Western capitalist society, and so sociobiology reinforced the "naturalness" of patriarchy. It also "explained," and so condoned, the existence of violence in society. The latter, coupled with a belief in genetically determined race differences in IQ, allowed sociobiologists to sanction racism as well as sexism, either intentionally or unwittingly. Their ideas have been incorporated into the ideology of the New Right, and as such they have been used as powerful political weapons, not only by right-wing extremist groups such as the neo-Nazis, but they have also gained electoral middle ground and even been carried into national elections (e.g., Hader in Austria and Le Pen in France).

Not coincidentally, the increased interest in biological determinism of sex differences in behavior began when the new feminist movements were making major gains. One of the key attractions of these neobiological social theories is its implied legitimation of small government and of social control (a key right-wing agenda). Any claims that governments might have a duty to address social ills, gender injustices, and poverty can ultimately be dismissed as irrelevant constructionist sentimentalities. If homelessness and crime are preprogrammed in the genes it is not a matter of offering help or rehabilitation but of social control. If men and women merely act in accordance with their biology, then their respective places in society are "natural" (i.e., nature-given) and, in some beliefs, even God-given. Any protest by women against their social roles must therefore be unnatural as would be policies to address gender injustices in social practice. By purporting to explain the structural divisions within society as natural, these theories have been used to reconfirm the status quo, and to argue that to work toward equality between the sexes, races, or classes is futile. One is reminded here of Max Weber's *The Protestant Ethic and the Spirit of Capitalism* (written in first English translation 1930) insofar as he outlined Calvinist views of success as a sign of grace and lack of success as a sign of divine wrath. The poor or the infirm were themselves to blame for their condition and hence undeserving; injustices were God-given.

Although some feminists took up the gauntlet of sociobiology quite early (e.g., Birke 1982, Bleier 1984, Rogers and Walsh 1982, Sayers 1982) most either felt that their knowledge of biology was insufficient to tackle the issues head on, or they were unaware of the importance that such biological determinist theories would attain. Moreover, in the debates on biological determinism versus sex differences, when and where they occurred, feminists failed to arrive at a coherent platform, and even now they are still divided on these issues (to be discussed later). Still, Sarah Hoagland (1982), among other feminists, forcefully argued at the time that biologically informed norms and beliefs about women are insidious because they prevent any form of resistance. Deviation from the norm, as long as that norm is seen as a natural given, must be abnormal and undesirable and cannot ever be recognized as resistance, revolt, or protest. The more women gained in social status and in independence through the 1970s to the 1990s, the closer the noose was tied in biological theory building against such freedoms. Sociobiology, and later evolutionary psychology in the lead, developed theories as if designed to regain men some exclusive social rights, more sexual freedom, and, ultimately, win the highest prize: unchallengeable biological supremacy.

For instance, in the 1980s, a number of sociobiologists formulated genetic models by which they claimed to be able to explain the mating patterns of both human and nonhuman species (see Austin and Short 1984). On a "cost–benefit" basis, the female is said to invest a greater biological cost in reproduction than does the male. As she must bear and nurture the offspring, her costs are best benefited by obtaining a male partner and making sure that he stays with her to assist in raising the offspring. On the other hand, for males, reproduction has very little biological cost and the best strategy for a male is to impregnate as many females as possible, thus maximizing the passing on of his genes.

The cost–benefit analysis has compelled sociobiologists to deduce that monogamy for women and polygamy for men are biological imperatives. Therefore, they say, these mating patterns are genetically determined and, by implication, they should not be disturbed by changing social values. Applied to human behavior, as it has been by evolutionary psychologists, this theory resurrects support for the philandering husband and even implies that the laws against polygamy are biologically wrong and hence unfair. Thornhill and Palmer (2000) have taken the theory so far as to claim that rape is a biologically adaptive strategy by which men, especially those who are unsuccessful in obtaining a partner, can nevertheless propagate their genes into the next generation. According to its followers, the theory explains why men rape and women rarely do so. Of course, it does much more than attempt to explain it;

it gives rapists some biological validity in a way that some may wish to interpret as tacitly condoning rape. Alternatively, it provides government agencies with a lever to reintroduce compulsary chemical castration as the only way perceived to solve the problem of offenders.

To clinch the argument sociobiologists claim that differences between women and men stem from genetic selection by evolutionary processes during our hunter-gatherer past (e.g., see Barkow et al. 1992). Earlier, biologically motivated arguments of "natural" differences between women and men were based not just on women's reproductive status but also on the female role, allegedly persisting over the past four million years. The argument, first elaborated by Tiger and Fox in 1972, assumes that prehistoric life of hominids and *Homo sapiens* fell neatly into gender roles (women as gatherers and men as hunters) and that this role difference was based on genetic and hormonal differences. The aggressive activity of hunting was apparently exclusively the domain of men while women reproduced; this allegedly explains why men are political leaders and women are caregivers today. Although such simplistic views (compressing complex processes over inconceivable time spans) have long been discredited (Barber 1994, Leakey 1994), they have reemerged in evolutionary psychology as concepts of the Stone-Age brain. According to this view, expounded by John Tooby and Leda Cosmides (1992), our evolution as human beings was arrested in the Stone Age and we have since been locked into our ancient biological selves. According to these writers, we have evolved immutable, unchanged gender roles dictated by our genes.

The upshot of such views, as they travel from the scientific and academic circles into general opinion and political fora, is that they give license for policy change in a backward direction. On the assumption that Stone-Age brains have genetically fixed sex roles, any infrastructure designed with social equality in mind must therefore be false. Help and assistance for women and a measure of freedom to control their reproduction then become matters of "social engineering" (derogatory meaning to prevent reform), rather than reflections of a just and democratic society of the twenty-first century.

Once such views translate into political debate and policy, it is typical that they encourage greater expression of similar views until the submerged network of hidden opinion by a few (Melucci 1988) acquires semblances of normality and reasonableness. And herein lies its greatest social and political danger, quite apart from its serious intellectual failings. In this climate, the 1990s witnessed a renewed assertiveness of the "rightfulness" of discrimination within the framework of a renewed interventionist approach to population control. These deliberate discriminations have manifested themselves not just in the literature but also in new population policies (Winant 1997). Under President George Bush, for instance, family planning clinics were prohibited from providing information about abortion (Tatalovich 1997) and a consistent assault on women's support structures, family clinics, and overseas aid (associated with abortion information in family clinics) was waged. The battle against women (Solinger 1998) has continued even more vigorously and extremely since the inauguration of his son George W. Bush into the oval office in 2001. In the 1970s, public proposals for policies to decide who should and should not reproduce would have been dismissed as far-fetched and, rightly, not taken seriously. In the climate of opinion created in the 1990s, one can no longer be so sure of this. Jarred Taylor's (1992) book concludes with a chapter on "Ending Reckless Procreation," in which he abandons the idea of debating population control *at the level of population* and instead proceeds to outline whose births should be prevented on an individual basis. In his opinion, the "underclass" should stop reproducing. He suggests that welfare payments should be made only to those women who agree to accept an implanted contraceptive known as Norplant, and that a government agency should be empowered to enforce rules of contraception selectively, as it sees fit. In the American debate, the issue of class is never far away from the issue of race (Herrnstein and Murray 1994, Steinberg 1995). In 1999, an abridged version of J. Philippe Rushton's book *Race, Evolution and Behavior* appeared, making a renewed attempt to restore Eysenck to his "rightful" position, in Rushton's opinion, as one of the greatest psychologists of the twentieth century. Rushton likewise reaffirmed Herrnstein and Murray's view (1994) that black Americans were inferior to white Americans and, in addition, suffered a number of *genetic* flaws. If the two arguments are put together (enforcing rules of contraception and the inferiority of black Americans), it is not difficult to see the potential makings of a new and viscious form of eugenics and the direct political control of reproduction.

The new vigor of male-driven attempts to control women's fertility and eliminate control over their bodies has made great inroads into society. It is not just a fight of male versus female, of power over reproduction, but of deterministic claims that biological differences between the sexes are fixed and hence the battle for reassertion of control is nothing but a step in the direction to re-create a "natural" biological equilibrium.

Biological determinist theories for sex differences in behavior are, of course, not new (Salzman 1977, Sayers 1982). Biologism dates back to the last century (Rose 1976), but over recent years the theories have been revamped into somewhat different forms. In the case of sex differences in behavior, it is no longer claimed that women lack the area of the brain concerned with intellect but, rather, that female brains are less asymmetrically organized than male brains and that this is determined by genes and hormones (e.g., Kimura 1999). Theories of this kind encourage and develop paradigms of simplistic dichotomies. Ultimately, political beliefs can hold greater sway than scientific commitment, allowing simple polarities to flourish in the absence of any evidence (Barkan 1992), be this in discussions on race or in debates about alleged dichotomies between femininity and masculinity. "Masculine" values have often been employed to justify imperialist expansionism and racism, people of other races being viewed as "soft" and feminine, less able to look after themselves (Birke 1982, Hoch 1979).

History has shown so far that, at times, genetic theories for human behavior have been an effective, though morally appalling, strategy to maintain power, but they are not based on good scientific evidence. The substantial scientific advances in genetics have increased public awareness of human biology and have inadvertently provided a receptive ground for more far-fetched biological explanations and political proposals in line with these explanations.

SEX AND GENDER

Some women refuted the claims of sociobiology directly, while others began to examine systematically their social and biological roles. The women's movement of the early 1970s had a long list of grievances and urgent matters to address. Inevitably, feminists themselves asked whether there was something "essential" about being a woman or whether the accepted social roles of women were largely a construct and far removed from the reproductive role of women. Sociobiologists, and now evolutionary psychologists, appear to have been generally under the impression, as far as one can glean from their writings, that all feminists are "constructionists" and deny a basis in biology (e.g., Alcock 2001). This is a misconception. From the start, feminists acknowledged some biological basis of difference and this they termed *sex*. In distinction to their own female sex, they also began to use the term *gender*. In its most fundamental aspect, sex refers to the biological difference between man and woman and is meant to suggest no more than a reproductive (beget/bear) difference. Gender, on the other hand, is a sociohistorical construction from which a vast number of social, political, economic, and legal consequences flow. To some, gender signals the translation of biological facts into social meaning (Scott 1988). But these two terms are not the same and are not interchangeable. Clearly, the agenda for political change requires close attention to "gender" roles rather than to "sex" roles. It is further a gross misconception of evolutionary psychologists, who presumably do not read widely in this field, to think that women, even when called feminists, all adhere to any agreed position on biology. There are essentialists among feminists as well as nonessentialists. In these chapters we want to briefly outline what this means and how it has informed debate.

The implications of the inferred differences between the terms *sex* and *gender* are profound. If one argues that woman is essentially different from man, the starting point of the debate must be that this difference ought to find recognition in the public sphere. Consequently, laws and policies ought to reflect women's special virtues and strengths (Ross 1965, 1977). The writers and proponents of theories of sexual difference vary from essentialist insistence on women being fundamentally different from men, to celebrating women's culture as something distinct. At times, they do this without clarifying their theoretical position to biology and essentialism. The French *écriture féminine* movement has at some times been regarded as essentialist and not at other times. In the sexual difference debate the concept of "universal citizenship" is rejected. According to Heller (1982), women may even require a specific form of polity, suggesting that women spontaneously reject submission to power and consequently tend to desire to live in a society without domination.

This distinction between sex and gender is specific to English-language feminism. For linguistic reasons it cannot be made in other languages, or not as well, and, to our knowledge, it has never been made in French or Italian feminism. Even in English-speaking countries, the gender–sex distinction is in some difficulty, or at least the term *gender* is often regarded as problematic. As Haraway (1991) pointed out, gender is an opaque term taken from biology and linguistics, gaining little clarification in the process of that adoption (Butler 1990, Snitow 1991). The main impetus for a critique of the term *gender*, and with it the gender theorists, came from semiotics and from those writing about sexual difference, foremost among them the French writers, including Helene Cixous (1974, 1975, 1986, 1987) and Lucy Irigaray (1974, 1977, 1984). Hence, internationally speaking, a division, albeit not a neat and uniform one, occurred between French (sexual difference) feminism (Gross 1986, Whitford 1991) and Anglo-Celtic (gender) feminism. Italian feminist thought has largely ignored the machinations of Anglo-feminism, but this is another story that has been told elsewhere (Bono and Kemp 1991, Kaplan 1992, 1993). Generally, those who use the term *sex differences* as a general description of women tend to pursue a politics of difference, while those who are focused on social and political change tend to use *gender* exclusively.

The term *sexual difference* is also not necessarily used in an essentialist or reductionist way. Some feminist critics of *gender* do not wish to use it because they believe that it fails to address the specificity of current cultural/social/political/economic "femaleness." The critique is thus largely situated in culture but, in some cases, the link is pronounced in an oblique way. Gatens (1991), a strong defender of the French *écriture feminine* movement, has perhaps given one of the most pertinent objections to the pursuit of a politics of sameness or "universal" political citizenship by doubting that a person can be split into a sexed body and an unsexed mind. Indeed, the old body/mind distinction in philosophy has not entirely vanished in the sex/gender distinction.

Biologists usually use the terms pragmatically: *sex* for physical/reproductive differences and *gender*, if it is used at all, for behavior. This is the way we have chosen to use the terms in this book. The gender/sex split in terminology and the uneasy use with which either term is occasionally employed in feminist writing well reflect the difficulties of those debates and, on the part of many feminists, the reluctance to commit entirely to rigid viewpoints. There are shades of gray between biologically determined and flexible aspects of the differences between women and men, and if scientists and social scientists cannot even agree to what extent men and women are really different, why should feminists? Indeed, the theoretical minefield in feminism may reflect the recognition of the complex interactions among genes, hormones, and the psychosocial environment. Before discussing this interaction in more detail, as we do so later, we will give a brief introduction to genes, hormones, and experience.

3

Limits of Biological Explanations

We begin this chapter with a brief overview of how the genes are expressed in the cell and how they relate to hormones. We consider this important to do because an understanding of the biological processes involved underpins our discussion of the psychological and social implications of research in these areas.

ROLE OF GENES

The genes, made up of DNA, are located inside the nucleus of each cell in the body and are passed on from one generation to the next in the egg and sperm. Some DNA is also passed from mothers to their offspring on the mitochondria, which are organelles in the cytoplasm of the cells. Each gene has a code for making a particular strand of messenger RNA by a process called transcription and that, in turn, makes a particular protein by a process called translation. A single gene, for example, codes a protein that influences eye color and another one does the same for hair color. The code is much more complicated for other physical features and for the structure and functions of parts of the brain. Not surprisingly then, there are many thousands of genes, about thirty thousand in humans, strung together into chains called chromosomes, but not all of the genes are expressed at any one time and, in fact, some are never expressed in a lifetime. Each cell in the human body has 46 chromosomes that come in pairs matched for shape, apart from one pair of particular relevance to us here. These two are the X and Y chromosomes, the Y being the smallest of all the chromosomes. These are also known as the sex chromosomes because they influence the development of female and male physical characteristics. Women have two X chromosomes and men have one X and one Y chromosome. A gene called SRY on the Y chromosome turns on (or activates) many other genes that then cause the testes to develop (Greenfield and Koopman 1996).

Genes located on the X and Y chromosomes are often assumed to influence the development of sex differences in cognition and behavior. They are said to exert their effects by way of the sex hormones (testosterone, estrogen, and progesterone in particular, but there are also other related hormones). A chain of influence is seen to act from genes to hormones to brain structure and function and then to behavior. This is often the only chain of influence that is considered at all, even though there is an equally important one in the reverse direction, from behavior to hormones to turning on the expression of certain genes. In other words, behaving in certain ways can affect hormone levels, altering the pattern of genes that are expressed. All individuals share and express many of the same genes but there is variation in the expression of some of the genes and in the degree to which each one is expressed. Furthermore, the pattern of gene expression changes with age and with the influence of factors in the environment. Sociocultural factors are part of this environment, and indeed they are a major aspect of the "experience" that we discuss later.

The mere existence of biological and behavioral differences between individuals or groups of individuals (such as women and men) does not prove that only the genes have determined these differences in a unitary way without any influence of experience. Moreover, the behavioral differences between the sexes are neither as absolute nor as common as both scientific and nonscientific reporting have given us to believe. Even the physical differences between males and females are not absolute (Kaplan and Rogers 1984, Rogers 1981). We discuss similarities between the sexes further in Chapters 6 and 7.

GENES AND HORMONES

The X and Y genes determine whether the fetus will develop ovaries or testes, which secrete the sex hormones. We refer to testosterone as a male sex hormone because, on average, it is present at higher levels in the blood of males compared to females, and we refer to estrogen and progesterone as female sex hormones because they are higher in females than in males, especially during pregnancy. But the dichotomy of male versus female sex hormones does not describe the biological condition adequately. Males also secrete the so-called female sex hormone, estrogen, and females secrete the male sex hormone, testosterone. Females secrete testosterone from the adrenal glands and not the gonads. Indeed, quite a sizeable percentage of females has higher plasma levels of testosterone than does another sizeable percentage of males.

In the brain, where sex hormones are meant to act to cause sex differences in behavior, the differences between the sexes are even less distinct. Most of the research in this area has been carried out using rats and mice as models since they are readily available laboratory species and the work requires killing the animals and examining their brains. Rodents also exhibit male–female differences in sexual behavior and thus brain-behavior associations can be investigated.

The first important point to note about the way the sex hormones affect organs is that they must bind to specific receptors in order to have an action. Therefore, those tissues on which the hormones act have receptors for them; this includes the primary sexual organs (the genitalia), the secondary sexual organs, and the brain. There are different receptors for testosterone, estrogen, progesterone, and a number of other sex steroids, and the number of each type of receptor varies from one organ to another. Certain parts of the brain have more receptors than others for the particular steroids. For example, it is known that the medial preoptic region of the rat brain contains more estrogen receptors than any other brain region (Herbison and Theodosis 1992). But is there a sex difference in the number of these receptors? The evidence collected so far suggests that there might be a sex difference of this kind (Brown et al. 1992) but it is nothing like as great as in the gonads and in other tissues where hormones act in the body.

If we delve further into the cellular basis of brain function, we find the division between the sexes becomes even more minimal. It is well known, from work on rats, that testosterone must be converted to estrogen inside the cells before it can have any effect on their function (Hutchison and Steimer 1984). In other words, the male sex hormone has to be converted to a female sex hormone to have any action in the male rat brain. Recent evidence, however, suggests that this conversion of testosterone to estrogen may be less important in the human brain (Cooke et al. 1998).

An aromatase enzyme carries out the conversion of testosterone to estrogen in the brain. The level of this enzyme in certain areas of the brain can be altered by social stimuli related to sexual behavior. Thus, there is no direct, one-way influence of the hormones on the brain; environmental stimuli can influence the power of these hormones to act. There is an intimate interweaving of internal and external input in the complex, whole system. To complicate matters further, at a cellular level, while the tissues outside the brain (e.g., the gonads) may recognize testosterone as testosterone and estrogen as estrogen, this distinction is less relevant to brain function. It is not even known whether a distinction between estrogen and testosterone is relevant to nerve cells, particularly as far as controlling behavior goes. This point is illustrated clearly by an experiment conducted by Nordeen and Yahr (1982). They raised the level of estrogen in the preoptic region of the hypothalamus in the right side of the female rat's brain by implanting a very small amount of the hormone there and found that this led the rat to adopt the mounting posture typical of male sexual behavior. The same treatment on the left side had no effect on mounting behavior but, instead, it suppressed lordosis, the posture typical of female sexual behavior, by about one third. Not only does this show the variation of the hormone's effect within the brain but also that the rat's brain has the nerve cell circuits for both female and male sexual behavior.

The more we discover about the way the sex hormones act within the brain, the more blurred becomes the distinction between the sexes. Since sex differences in behavior exist, it has been inferred that the structure and function of male and female brains is different, and that this is genetically and/or hormonally determined. We now know that this polarization of male and female does not apply in any simple way, if at all. This is not to say that the sex hormones have no effect on brain structure and function. There is plenty of evidence that they do. The point is that their various actions are not simple and distinct in males and females, nor do they act in isolation for interactive influences from experience and learning, as we will discuss in Chapter 7.

The actions of sex hormones on nerve cells are various and widespread, especially while the nerve cells are growing or developing. They affect the formation of new nerve cells and also their death; nerve cell migration from one region of the brain to another; the number of connections between nerve cells; and the expression of certain chemicals, called neuropeptides, which are involved in transmission of activity from one nerve cell to another. Their effect is greatest when the brain is developing, but they do have some effect on the structure of the adult brain, as recently shown to be the case for the brain region known as the medial amygdala: in the adult rat, the size of this brain region shrinks after castration (i.e., when testosterone levels decline) (Cooke et al. 1999).

Without question, sex differences are present in both brain structure and behavior in humans and animals. But the evidence for such differences is much stronger for nonhuman species than for humans. For example, in rats, some regions differ in size and other regions, such as the hippocampus, have different numbers of connections between nerve cells and different numbers of receptors for neurotransmitters (reviewed in Choleris and Kavaliers 1999, Cooke at al. 1998). There is no evidence of anything similar in humans, although it would be much more difficult to find out in humans (more later). There have been experiments showing that sex hormones can influence the development of neurons and alter the organization of the brain. Nevertheless, the same patterns of neuronal growth are also influenced by environmental stimulation, brain injury, nutrition and, in some regions of the brain, the action of other hormones such as thyroxin or corticosterone. A plethora of influences impinge on the development of each neuron and even at the cellular level, it is impossible to single out a unitary influence of sex hormones as the possible starting point for causing sex differences in behavior. All too often researchers have interpreted their results as showing sole and direct effects of the sex hormones on the brain without considering that they might affect brain development indirectly by changing the social environment in which the individual is raised. We will now discuss the most important example of this.

In experiments conducted in the 1960s, using rats exposed to testosterone during the first five days after birth, it was found that in later life, sexual behavior typical of the male (i.e., mounting behavior) was exhibited irrespective of the rat's genetic sex but dependent on whether testosterone levels had been high in the first five days of life (Harris and Levine 1965, Phoenix et al. 1968). This result was said to show an organizing effect of testosterone on the brain. Castrated males and females not exposed to testosterone in early life were found to develop sexual behavior typical of the female (i.e., adopting the crouching, lordosis posture during mating). It appeared that the presence of testosterone in neonatal life directed the brain to develop in a typical male-type direction, and that in its absence a typical female-type brain developed. Testosterone thus appeared actively to operate a switch mechanism channeling brain differentiation toward maleness, and the female brain was a default state, one developing in the absence of testosterone.

This apparent "fact" about the active effect of testosterone in the development of the male-type brain and the default condition of the female-type brain was widely accepted, and still is, even though two lines of evidence show that it no longer applies. Recent evidence shows that the default condition for females is incorrect, since the female sex hormones can actively affect brain development in their own right (Bimonte et al. 2000, Fitch and Denenberg 1998). Furthermore, as we discuss in the next section, testosterone organizes the brain to develop in a male direction not directly by acting on the developing brain but indirectly by changing the behavior of the mother toward her pups.

Here we need to say that the above work on rats and the early misinterpretation of it has been extremely influential on theories of sexual development in humans. Despite the fact that there is no evidence for long-lasting effects of early androgen exposure in primates, researchers influential in the field of sexology (Dörner 1979, Money and Ehrhardt 1972) extrapolated to humans the data obtained using rats (see Rogers 1981). Money and Dörner went so far as to suggest that male homosexuality is caused by insufficient exposure of the genetic male to androgen in fetal life, and lesbianism is due to excess exposure to androgen of the genetic female. The work of Money and Dörner has been criticized extensively elsewhere (Birke 1979, Rogers and Walsh 1982). The essential point to consider here is that they ignored the importance of experience and the interaction between sex hormone levels and experience in leading to femaleness and maleness.

It is also important to note that we mention this research, now two or three decades old, because it was so influential on thinking from then on, and remains so (e.g., Geschwind and Galaburda 1987, Hamer and Copeland 1998, Hines 1993, Le Vay 1993).

GENES AND EXPERIENCE

Relatively few studies have investigated the effects of experience on the differences between the sexes. It is easier to conduct controlled research on animals than on humans, but to even consider that it might be interesting to investigate the effects of experience on the development of sex differences in animals presupposes the insight that parents might treat their male and female offspring differently. Until recently, few researchers thought of this as a possibility. The dominant mode of thinking about animals accorded relatively little influence to learning and it followed that sex differences in behavior were regarded as controlled largely by the genes and sex hormones. Although some researchers in the field did recognize an effect of learning on sex differences in humans, fewer extended this possibility to animals and this remains true today.

If experience does cause males and females to develop and behave differently, that experience, in the first instance, would have to be contained in the differential treatment of males and females by other members of their species. For rats kept in laboratory conditions, the mother is often the only adult with whom the pups have contact. Does the mother treat male and female pups differently? It is now known from a number of studies, and for more than one strain of rat that the mother rat tends to her pups according to sex. For instance, it was found that she licks the region around the anus and genitals (referred to as anogenital licking) of male pups more than she does the same region in female pups (Moore and Morelli 1979, Moore et al. 1997). The surprising finding was that the discriminating behavior of the mother toward male and female pups affects the development of sex differences in both brain *structure* and *function*. In fact, the differential experience received by males and females overrides any influence of genes or sex hormones on sex differences in behavior, as shown by an important series of experiments conducted by Celia Moore and colleagues (Moore 1982, 1984, 1995).

One of the experiments conducted by Moore showed the importance of the early experience of anogenital licking by artificially increasing the stimulation of the anogenital region in females by stroking them with a paintbrush. Having received this additional stimulation as pups, the females behaved the same as males when they reached adulthood. Their activity patterns were the same as those of males. They mounted other rats in the copulation act, as do males, instead of adopting the female crouching position. They even secreted luteinizing hormone from the pituitary gland continuously, as do males, instead of showing the cyclic pattern characteristic of females. Therefore, both their biology and behavior had been changed to that typical of the male merely by stroking the anogenital region during development.

Other experiments showed that the mother distinguishes between her female and male pups by the smell of their urine (Moore 1981, 1985a). By inserting small tubes in the nasal cavity of the mother it was possible to prevent her from being able to smell the differences between her male and female pups. In the absence of appropriate olfactory information she licked the anogenital regions of male and female pups in equal amounts and no sex differences in behavior developed. Furthermore, castrated males, Moore found, received no more licking than females and they exhibited female patterns of behavior in adulthood. They also had the female-typical cyclic pattern of secretion of luteinizing hormone. It is, in fact, the presence of the male sex hormone, testosterone, in the urine of the pup that attracts the mother to lick the anogenital region of male pups more than female pups. This is because the testosterone causes a gland in the anal region to secrete an attractive substance. Female pups injected with testosterone are licked as much as male pups and they develop male-typical behavior and an acyclic pattern of secretion of luteinizing hormone. Hence, the sex hormone, testosterone, plays a role in the differentiation of sex differences in adult behavior but it acts indirectly by altering the mother's behavior and not by acting directly on the developing brain.

The experience of anogenital licking also stimulates development of the testes and increases the size of the sexually dimorphic nucleus, a region in the spinal cord, which is made up of nerve cells that innervate some of the muscles of the penis. The size of the nucleus increases as a result of the addition of more nerve cells during the growth phase (Moore et al. 1992). The anogenital stimulation increases the size of the nucleus in both males and females but the nucleus size in stimulated females is not as large as that of unstimulated males. In the case of this nucleus the sex hormones and the effects of experience interact to determine the size it will reach. This is in contrast to mutual effects of sex hormones and anogenital licking on sexually dimorphic behavior and secretion of luteinizing hormone since, in these cases, the effects of anogenital stimulation completely override the effects of testosterone. In natural parental care, of course, the effects of testosterone and experience interact.

As Moore and colleagues (1996) point out, sex differences in development might be initiated by different levels of secreted sex hormones in males and females, but the subsequent means by which sex differences develop are diverse and indirect. For example, within the central nervous system, some nuclei respond to sex hormones, others to stimulation of the anogenital region of the pup by the mother at different stages of development, and some to the interaction of both these contributions. There must be diverse effects on behavior. In other words, there is no unitary causal relationship between the sex hormones and sex differences in behavior. It follows, therefore, that the reductionist approach is an unsatisfactory and incorrect way of explaining sex differences, even in rats.

To summarize: these investigations have revealed the importance of experience in the development of sex differences in brain function. They demonstrate the interaction of genes, hormones, and experience in the development of sex differences. Similar interactive processes are bound to be involved in the development of sex differences in other species, including humans, although the type of experience that is effective may vary from species to species.

It is now clear that the earlier experiments, which were thought to have demonstrated that early androgen exposure of rats causes the brain to differentiate in a male-type direction due to the action of the male sex hormone on the brain, must be reinterpreted. The androgen (testosterone) treatment in early life was altering maternal behavior, and thus indirectly altering the development of behavior through external environmental stimuli, or a learning process. Researchers have been all too ready to assume that behavior patterns in animals are directly genetically and/or hormonally controlled. Moore's work has been extremely important. It clearly elucidates the fact that generally accepted views of the direct link between effects of early hormones on behavior through direct effects on brain differentiation are incorrect. The sex hormones co-act and interact with other factors throughout the individual's development.

In our opinion, this area of research is exciting, but nearly two decades since Moore carried out her original experiments with rats, only a few other species have been examined in similar ways and the work is not cited as widely as it should be. On the other hand, perhaps we should not be surprised by this lack of interest because research on the direct and unidirectional genetic and hormonal causes of sex differences has expanded, and reductionist approaches continue to hold sway in the field.

Nevertheless, a few studies along these lines have been conducted. A similar bias for mothers to spend more time licking the anogenital region of male offspring has been found in ferrets (Baum et al. 1996). Also, some research on parental behavior in rhesus macaque monkeys has been carried out and the picture emerging is that in this species, sons receive less maternal contact than daughters. In primates, social rank also plays a role in mother–infant interactions. One study of rhesus macaques reported that sons of high-ranking mothers are rejected (pushed away or ignored) more frequently than daughters of the same mothers (Simpson 1983). A study on free-ranging yellow baboons (Wasser and Wasser 1995) and another on rhesus macaques (Brown and Dixson 2000) also established that sons of low-ranking mothers were given less close physical contact with their mothers than were daughters. Long-tailed macaque mothers also groom and hold their daughters more than their sons (Nakamichi et al. 1990). Although these observations of differential treatment received by male and female offspring raise the distinct possibility of this influencing the development of sex differences in behavior, so far no research has addressed this topic directly.

Now we might ask how relevant the studies on sex differences in animals are to humans. Clearly, anogenital licking is not a behavior common to humans. The behavior in rats showed, however, that in a very important aspect of mother–pup interaction, the mother treated male and female pups differently and did so consistently. There were also measurable outcomes when this natural behavior by the mother was purposely intercepted by the experimenters. Do some of the same mechanisms of development exist in humans? What evidence is there that mothers and other family members treat boys and girls differently? There is, indeed, evidence that mothers encourage different play activities in boys and girls and select different toys that they see as appropriate for girls and boys (Campenni 1999). One of the classic and often cited studies in this area tested the responses of women to the same child dressed as a girl or as a boy (Will et al. 1976). When the child was dressed as a girl, the women spoke to "her" gently and encouraged "her" to play with dolls. When dressed as a boy, the child was encouraged to play with toys such as trucks and hammers. The adults were taking an active role in encouraging these young children, only a year old, to conform to the male–female stereotypes, although they were unaware that they were doing this.

With young boys, adults often employ coercion rather than encouragement. Researchers have found that adults are stricter with boys than with girls in ensuring that they play with toys considered to be acceptable for their gender and that fathers tend to differentiate between their sons and daughters more than do mothers (Lytton and Romney 1991). When they are just a few years older, as other studies have found, the children themselves select gender-specific toys and encourage these choices in their peers (Martin et al. 1995). They are also conscious of their parents' attitude to their choice of toys and play, and boys especially report that cross-gender-typed play is "bad" (Raag and Rackliff 1998).

In the above-mentioned experiment, the clothes in which the child was dressed signaled the sex of the child. As far as the child's parents are concerned, the structure of the child's genitalia determines whether they treat the child like a boy or a girl. A girl with an enlarged clitoris may be treated like a boy, and a boy with a small penis treated like a girl, unless the parents are informed about the genetic sex of the child (i.e., whether the child is XX or XY). This sex-limited treatment must provide ample differences in experience leading to the development of sex differences in behavior. Nevertheless, it is surprising how few studies have tested this experimentally, compared to the plethora of studies on the genetic and hormonal influences on sex differences in behavior.

Lytton and Romney (1991) examined the results of 172 studies that had investigated whether parents treat their sons and daughters differently. Their sample came from a number of cultures. Their meta-analysis revealed that there was little evidence of differential treatment for girls and boys in terms of some specific treatments, which included the amount of interaction between parent and child, applying discipline, encouraging dependency, and some other measures. However, they did find a significant effect of both parents on encouragement of sex-stereotyped play activities and household chores. Also, in Western countries apart from North America, boys were disciplined with physical punishment more commonly than were girls.

As in the various studies of mother–infant interaction in animals, the analysis of the large number of human studies demonstrates that we need to be very circumspect and carefully distinguish between the kinds of interactions being observed. On one level, there were no reported differences in treatment between male and female offspring. In crucial ways, however, the parental behaviors showed strong sex bias in their responses to the sons and daughters (play and household chores). Not all parental behaviors need to address sex differences in order for some to have a long-lasting effect on the behavior of offspring.

Evolutionary psychology usually disregards subtle interplays like this as marginal or insignificant. Campbell (2002) is no exception. She dismisses the results (of the same meta-analysis) and uses them as proof that there is little evidence for parental training of sex differences in behavior. Her own hypothesis is that sex differences are determined by the genes: that is, boys and girls are born with cognitive differences and they then make choices that further channel their behavior into masculine- and feminine-typical patterns. As further support she cites her own research showing that boys and girls choose to play with different toys as early as 9 months of age. For the following reasons, we do not agree with her interpretation of these two sets of results.

First, the study by Lytton and Romney (1991) did, in fact, show that parents treated their sons and daughters differently by encouraging them to play in ways considered desirable for their sex and to carry out sex-typed chores in the household, even though few other differences were detected. In our opinion, Campbell overlooks this important fact. Even small differentials caused by parental encouragement of sex stereotypes could lead to marked differences in behavior in adulthood, as Lytton and Romney agree. It is also possible that the researchers did not include other, perhaps more subtle, cues that reinforce different behavior patterns in boys and girls. Further, there are many events that studies of this kind do not address, such as a child's identification with a parent, trying to please a parent, power games that are being played in the family, gender-specific interactions (same sex identifications/role models), and many other psychological factors not measured in these studies. Methodologically, there are other considerations. For instance, the timing at which certain types of learning take place may be profoundly important. Experiments with animals have proven that as little as a matter of hours can have marked and long-lasting effects on behavior. One of the most dramatic effects occurs in imprinting in animals. This is a very rapid and powerful form of learning by which a young precocial animal (e.g., chick, duck, sheep, or goat) acquires, often within minutes, the ability to recognize its mother, as a matter of survival. Imprinting occurs during a sensitive period just after hatching or birth and it alters behavior from then on. To take another example of rapid learning in early life: research from our own laboratory has shown that raising young chicks in a group of ten, compared to pairs, for as little as twelve hours on the first day after hatching, completely alters their ability to choose between a cage mate and an unfamiliar chick when they are tested later on (Deng and Rogers 2002). These studies of animal models demonstrate learning processes that may be analogous to those in humans. We are not suggesting that research on animals proves anything about humans, but it raises questions and possibilities that should be investigated. Such rapid learning would have been missed in all of the studies that were part of the meta-analysis. All of these points lead us to disagree with Campbell's decision to ignore the significant finding of differential parental encouragement of play stereotypes in children.

Second, we address Campbell's claim about her own research on choice of toys by 9-month-old infants. Children learn before the age of 9 months, probably much more than psychologists have so far discovered (it is not easy to test very young children). There is no reason very young children would not respond to cultural constructs of gender, as manifested in choice of toys as early as 9 months. In fact, some time ago John Money showed that by as early as 1½ years of age children have developed such a strong sexual identity that, after they have reached this age, it is very difficult to get them to shift to the identity of the opposite sex (i.e., in cases of incorrect identification of the child's sex), although such a shift can be made if it is done prior to this age (summarized in Money and Ehrhardt 1972).

Our third objection to Campbell's conclusion is that social class or rank may have been an important factor, as shown in the macaque monkeys discussed earlier. Styles of human parent–child interactions are class dependent and, despite the fact that many aspects of sex differences in behavior cut across class barriers, the way in which these are learned may be very different according to where and with whom the child is growing up. There is much to be explored here before jumping to conclusions. The research on parental behavior of monkeys signals a need to factor in many of the same variables in studies of humans. Campbell seems to believe that these signals can be ignored but, surprisingly, seeks to confirm that evolutionary psychologists' explanations of sex differences are mindful of broader influences, as the following excerpt shows (Campbell 2002, p. 15):

> Where are these evolutionary psychologists who allege that genes operate without reference to hormones, experience or environment? Wherever they are, I have not been able to locate them. Rather it is environmentalists that [note: not "who"] continue to set up mythical distinctions between nature and nurture in order to maintain a clear line between the politically correct and incorrect.

Provided Campbell thinks that taking the environment into account means looking at a small glimmer of it at the end of a long tunnel, she is correct in saying that evolutionary psychologists take "it" into account. To evolutionary psychologists, the environment takes on an extremely limited, if not unitary, quality. To them, learning is strongly constrained by the genes. Our genes channel the environments we will choose and the stimuli to which we will pay attention so that our learning is channeled (or constrained) by the power of the genes. The beginning point for the process is, thus, located in the genes. They recognize that the environment influences what genes will be expressed subsequently but genes make us take the first step along life's trajectory and so determine the pathway of gene expression to follow. In other words, genes dominate. So, we agree, they do consider the environment but only in a very limited way. The concept of intermingling processes of development in a rich and varied context eludes them. They then accuse researchers who choose to take a range of variables into account in their studies of being "environmentalists." (Note: this has nothing to do with conservation but is meant to refer to important external influences such as learning and experience.)

There are sex differences in the human brain but we know next to nothing about the mechanisms by which they develop. If sex differences exist at or before birth, they cannot, of course, depend on differential treatment of boys and girls by the parents. In humans, there is only one convincing example of a sex difference in the brain at birth: the brain weight. On average, males weigh more than females at birth, and their brains are proportionally heavier (Pakkenberg and Gundersen 1997). This is due to the action of hormones prior to birth, probably an interaction between the sex hormones and the growth hormone. It is a general effect on all of the body and the brain and thus unlikely to be the cause of specific differences between the behavior of males and females. The only other reported sex difference at birth in humans is a sex difference in the size of the corpus callosum, a large tract of nerve cells connecting the left and right hemispheres of the brain, but this has been found in only one study (de Lacoste et al. 1986) and it remains controversial. The existence of a sex difference in the size of the corpus callosum is still disputed even in adult brains, where it has been studied more extensively. So far, it appears that sex differences in the human brain are far from being as obvious as those in the laboratory animals studied (reviewed by Cooke et al. 1998).

Based on the studies of rats demonstrating the interaction of experience and sex hormones in the development of sex differences, it is highly reasonable to expect to find such interactions in the development of sex differences in humans, although the exact nature of the interactions is likely to be specific to the species. Also, the kind of experiences that influence the development of sex differences might be more diverse in one species than another. We do not suggest extrapolating from the findings about rats directly to humans (as some researchers do), but this research does lay a basis for investigating these possibilities in humans.

As Moore (1985b) has so clearly outlined, we should not lose sight of the complexity of hormone–behavior interactions. Too many psychologists interested in sex differences have assumed that they stem from any organizing effect of the sex differences on the brain, and have ignored the interactive effects of experience. Surprisingly, they continue to do so even in the face of scientific evidence to the contrary. Moreover, to many, the mere existence of a sex difference in brain structure is thought to imply that the sex hormones have acted directly on the brain to cause the difference. This is a clear denial of the fact that experience and behaving in certain ways can alter not only behavior (brain function) but also brain structure. At the very least, it is disappointing to find that some psychologists persist in tying sex differences solely to the direct action of sex hormones on the developing brain, given the now vast amount of evidence showing the flexibility of the nervous system and the complex interactions between brain and behavior. We will discuss this further in the next chapter.

4

Biology and Flexibility

Biology is the study of life; this includes both the organism itself and the environment in which it functions. Yet biology and the environment are often spoken of as if they were at opposite poles. This thinking results from a flawed understanding of biological organisms and biological processes. The latter are neither static nor disconnected from the context in which the organism lives. Of course, any organism depends on its structure, its chemistry, and its physiology, but it does not do so in isolation from its environment. Genes are expressed in particular contexts and their very expression alters the organism's environment so that a new set of interactions takes place. In fact, so intertwined are these processes that we find it impossible to speak of them as separate entities.

What, therefore, should we do with genetic explanations of human or animal behavior? If we close the doors on explanations of behavior that lie outside the field of genetics, we would lose a good deal of our humanity. We would also lose a good many opportunities to understand the human condition now and in the future. We would lose our ability to see prisms of light and instead return to schemas of black and white.

Today we see enormous sums of money being put toward the Human Genome Project and other projects working on finding the genetic code of a handful of other strategically chosen species. By comparison, the funds for studying the effects of experience and learning on development are minuscule. We are under no illusion that there is any truth to the concept that science seeks truth in a vacuum outside the society in which it is practiced. The funding for research is far from solely driven by the challenge of new ideas, nor is it driven solely by a humanistic wish to make life better for the majority of the people (e.g., in terms of curing diseases). The latter is evident by the fact that far more people in the world die from malaria than by the expression of specific genes that cause particular syndromes, and yet relatively few funds are put into research on treatments for malaria. It seems to us and to others (e.g., Fausto-Sterling 2000, Rose 1995, Rose and Rose 2000) that the driving force for research is political gain for certain groups in the Western world. Beguiling as molecular genetics may seem in the present climate and promising as current discoveries may be, neither genetics nor evolutionary psychology can hope to hold the key to all that requires explaining.

There is always a temptation to grasp at all-embracing theories that might solve a specific puzzle or answer the problems of the world. Unfortunately, some scientists now are not solving the world's problems but are contributing to creating them. They create them theoretically, at laboratory benches, seemingly far away from the real world. They do so by anchoring biology and behavior to the genome, and they do so by failing to recognize that interaction and constant change characterize living organisms. As Steven Rose (1997) pointed out, the biologist intercepts living processes and freezes them at a moment of time (e.g., in a test tube or on a microscopic slide) and then forgets that the piece so captured comes from a dynamic and complex system. Experience and learning have always been less tangible to the scientist than have physical entities that can be trapped in a test tube or on a microscope slide. Flexibility and complexity have tended to be pushed aside in favor of oversimplified explanations. But what if the explanation we should be seeking is complex rather than simple? No amount of wearing blinkers will make the simple explanation match the full spread of evidence.

When biology enters the fray to explain human behavior, it seems to find itself confronted by an audience eager to snap up simple explanations (a gene for intelligence and another for aggression) and the players in this dangerous game receive more points (notoriety, book sales, etc.) if they give the audience what it wants, or what it has been slowly educated into wanting. The measured and circumspect manner of conducting science is drowned out in the clamor for sensational news, for popularity, and, regrettably, for political and monetary gain. But biology does not have to lend a hand in this process, as it has so often done in the past. To take this discussion back to sex differences, we can see that the renewed interest in genetic causation of sex differences was inevitable, given the enormous amount of funding for molecular genetics and the push to find genes controlling a range of human conditions. This new imperative has been superimposed on a much longer tradition of interest in sex differences. Let us consider what that means at the level of the genes and hormones.

FLEXIBILITY AND GENES

We all carry genetic material that is not expressed. In fact the majority of the DNA in our cells is merely junk accumulated from the past (possibly viral attacks, etc.). We still have an ample number of genes in each cell, although the Human Genome Project has told us the startling fact that we have fewer genes than the rice plant—this is a case of humans not being bigger and better. Even if we do not consider the genetic junk, among the genes that could be expressed only a subsection is expressed at any one moment of time and some will never be expressed. Gene expression is always changing in continuous interaction with the environment both inside the organism and surrounding it. Numerous kinds of experience and learning possibilities impinge on the organism, and the particular mixture of inputs influences exactly what cascading arrays of genes will be expressed. A change in the environment can "turn on" genes. There may be elements of chance and there may be genes that are expressed come what may, but the important point is that epigenetic factors get an open and interactive role to play in the process of life. It is, therefore, as important for biologists (and biopsychologists) to investigate epigenetic factors as it is to investigate the genetic ones. The environment is thus as important as the genome.

One of the least plausible and, at times, even bizarre elements of the new deterministic theories is to declare so many aspects of our human existence as "by-products," and to suggest that human behavior in its various expressions is as it is and is not really subject to moral evaluation, change, or improvement. Such is the bravado of evolutionary psychologists that some, at least, are prepared to take on the mind and declare it merely a by-product of the genes and their biological imperative to replicate and be passed on to the next generation. As Steven Pinker (1997, p. 43) says in his book *How the Mind Works*, "The ultimate goal that the mind is designed to attain is maximizing the number of genes that created it." According to him, the genes are the benefactors of our physical and mental design. The genes are therefore seen as the driving force behind all human endeavor.

Concerning sexual behavior, Pinker postulates that genes determine monogamy (in females) and polygamy (in males), aggression, and the perception of beauty. This notion might please a great many men. Let us imagine just for a flippant moment, a world organized around the "principles" that are uncovered here. Aggression (whatever its meaning), from now on, whether or not resulting in murder, will be treated leniently, as long as a male has committed the act either because he could not help it or because he has a "genetic right" to it. Such laws currently exist in some parts of the world, and in at least one country, suspected infidelity of a wife has been punished by cutting off her nose, breast, ears, hand, or cutting out her tongue, indeed, by blinding her or putting her to death. Let us imagine also that every man can take as many wives as he likes and the laws against polygamy are abolished tomorrow or, if he is unmarried, he can rape any woman he likes (he "needs" to do this because of his genetically programmed goal of passing on his genes). There are countries that allow polygamy and some in which it is a sign of prestige. But some of these social organizations have their roots in women's death during childbirth. To take on a woman was a risk because she might leave the father on his own with children. The dowry was developed to persuade men to take a wife, or even to build his wealth by having many wives. Imagine, if Western banks were to close because wealth accumulation and investment could now run once again on the dowry system. In some countries, some of these ideas are partially in force but there are usually many checks and balances. Obviously, such hypotheses must be meant for entertainment or for science fiction writers. If we took them seriously, a new Dark Age would be ours for the taking. Of course, if anybody thought of putting Pinker's theories into practice (and there are always some who will think of its applicability), most women in the Western world would want to apply for a sex change at once.

But here it is our concern to discuss the flexibility of biology. In the current climate replete with unsubstantiated claims of new discoveries of genes for this and genes for that, it is important to remind ourselves that genes do not operate in the simple manner of determining the physical nature of behavioral features of an organism in the sense that the direction of causation is from genes to behavior. They are but one part of the interactive system that characterizes life. We can discuss the ways in which genes are expressed inside cells at one level and, at another level, we can discuss the living being's psychology or social interactions. These are different translations of the process of life, and are all part of a whole that is more than the sum of its parts.

BRAIN/MIND MODULES AND THEIR PROBLEMS

More recent concepts in evolutionary psychology also have relevance to the debate on sex differences in behavior and even sexual orientation. Here we refer to the concept of modules.

The evolutionary psychologist considers that if genes are packets of information, then perhaps the brain/mind is built on packets or modules. This is relevant to our discussion because if so, discrete modules might determine those aspects of behavior typical of males while other modules determine those typical of females. Evolutionary psychologists also refer to them as specialized intelligences and see each as being dedicated to a specific domain of behavior, and each functioning largely independently of the others (Buss 1994, 1999, Fodor 1983, Hirschfeld and Gelman 1994, Mithen 1996). These modules are considered to be units of function, not necessarily located in discrete structures or regions in the brain. For example, there may be a module that processes the information that allows us to see faces, and another to see and make sense of objects, or one module might process the meaning of words and another the syntax or grammatical constructions of whole sentences, or even parts of them (Pinker 1994). These collections of modules, they suggest, might differ in males and females.

The function of the brain is said to depend on connections and interactions between these modules. Consistent with this hypothesis, there are people who seem to show specific defects in only one modular-like function, such as the stroke patient who can no longer discriminate between faces but can still identify objects (for further examples see Bradshaw and Mattingley 1995, Kaplan-Solms and Solms 2002). The functional lesion appears to be very specific in these cases.

Nevertheless, the modular concept of the mind has several problems. The first is the liability of modules to slip into assumed structural entities and, in turn, to be seen as building blocks determined by the genetic program. Once psychologists draw diagrams with labeled boxes designating modules (a box for object identification, another for identification of faces, etc.) connected to each other in various ways it is a short but treacherous step from this model to assuming that each box (or module) corresponds to an actually existing structure and has a genetic basis. Another such step then says that these genes and their modules are adaptations to the environment in which our ancestors lived and survived—with clear application to sex differences, since the other assumption is that the society of ancestral humans was divided strongly along sexual lines. Our former point is illustrated by the following quotation from Duchaine and colleagues 2001:

> The human brain is a set of computational machines, each of which is designed by natural selection to solve adaptive problems faced by our hunter–gatherer ancestors. These machines are adaptive specializations: systems equipped with design features that are organized such that they solve an ancestral problem reliably, economically and efficiently. [p. 225]

In addition to our objection to this mechanistic description of the human brain/mind, we see other reasons why these assumptions are problematic. Even if modular processing occurs, at least at some levels of cognition, and there are localized regions of the brain designated to a discrete function, there is absolutely no reason this organization should be genetically caused rather than being the result of experience or a variety of learning processes. Learning itself may generate modes by which cognitive functions develop (Karmiloff-Smith 2000). Since the existence of modular ways of processing tells us nothing about what caused them, there are no grounds to assume genetic causation.

The second problem follows on directly: modules may have limited applicability to brain function. Processing of perceptual information may be modular but modularity may not apply to higher cognitive processes, which are important in many of the sex/gender differences. We know from neuroanatomical and electrophysiological studies that different types of visual information are processed, at least initially, in parallel in different nerve pathways and brain regions (e.g., color, movement, and depth of visual stimuli are processed in this way). We can consider this to be modular processing, although it, too, has been questioned, since it is a matter of defining how discrete a module has to be (Holcomb 2001). Higher-level functions require integration of these perceptual inputs; here processing becomes much more flexible, to the extent that it could not be considered as modular.

We are not simply dealing with a semantic problem here or hair splitting on whether something is interconnected or modular. This is a very important issue and many assumptions follow from it. First, once modules as discrete information-processing capsules were "invented," the step was made to believing that they are under strong genetic control (often referred to as canalization of the genes). Then, once this assumed gene-module connection had been cemented, the evolutionary psychologists could afford to soften their position to appear more reasonable, and indeed to attempt to accommodate known neuroscience, while not perturbing the postulated gene-module deadlock. In a similar manner, Buss (1999), Cosmides and Tooby (2000), and other evolutionary psychologists might want to convince us that their hypotheses are not rigid, since any particular gene-dependent module is sensitive to context and not expressed in an invariant way. This is merely *post hoc* posturing and still leaves genes with an overriding role.

Although in their original conception modules were discrete entities operating largely independently of each other (see Fodor 1983), Cosmides and Tooby (2000) now hypothesize that the "multimodular mind" is made up of a collection of interconnected modules. Hence, they have taken us into a gray area, and the problem with gray areas in science is that they become untestable. The new modular theories have a foot in both camps (the brain is modular but it is interconnected and perhaps therefore nonmodular). Obviously, whether modules have any meaning at all depends on just to what degree they are interconnected. A completely interconnected system would not be modular.

Let us take a recent example in which modular thinking led to consideration of a single gene as the determining principle for language. Lai and colleagues (2001) reported that they had found a genetic abnormality in a British family, some of whose members were unable to produce speech and language. This finding was immediately taken up by evolutionary psychologists, anthropologists, and the media to claim that the gene for language/grammar had been discovered. Pinker (2001) said that now we might have a basis to go about finding out when that particular gene first appeared in early humans and then we would know when language evolved. What a simple answer to a vast question that has engaged us for a very long time! The problem with this simple solution is that the gene was associated with a specific defect in language that is likely to be quite separate from anything related to language as it normally occurs, which, incidentally, requires learning.

We should now think about postulated modules and sex differences. Perhaps no more needs to be said than the following: Different patterns of behavior mean, to the evolutionary psychologist, that there must be different modules present or different connecting links between the modules. By extrapolation, this difference must depend on fundamental differences between the male and female genome selected in our evolutionary past. But this line of thought ignores the known facts that learning and experience affect brain development in continuous interactive processes with gene expression; the mere existence of functional and structural differences gives us no clue as to how they came about.

NERVE CELLS AND FLEXIBILITY

To comprehend the biological processes important to understanding sex differences, it is also relevant to discuss some of the latest information on how nerve cells differentiate and function because this information, above all, reveals their flexibility. It is these cellular processes that evolutionary psychologists either ignore or actively avoid since they do not fit well with their rigid concepts of brain function.

The interaction of experience and gene expression is manifested in the behavior of the whole organism; it is also seen clearly in the nerve cells of that organism. A very large number of studies have shown that nerve cell growth and differentiation requires the expression of a changing array of genes. A new technique called serial analysis of gene expression (SAGE) is showing how the hormone, Nerve Growth Factor (NGF), affects the expression of one set of genes, then another and another so that the numbers of genes expressed form a changing cascade (e.g., Greene et al. 2001). The end result of the timed expression of thousands of genes within one type of nerve cell promotes its differentiation into a new form. The entire shape of the cell changes; it often migrates to a new place in the brain; it forms new connections (synapses) with other nerve cells, which reciprocate; and later it prunes some of these connections, usually in response to stimulation received from the environment. These are but a few of the known processes of nerve cell development and also of the changes that can occur even in mature nerve cells.

Neuroscientists studying these processes are amazed at their complexity even within a single cell. From this perspective, it would be nonsense to think of the expression of a single gene encoding a complex function, such as the rules of grammar or the motivation to rape (as per Pinker 1994, 1997). Neuroscientists are finally beginning to understand some aspects of brain function and to appreciate its enormous complexity when some theorists from psychology come along and think it is all so easy.

LEARNING AND EXPERIENCE

The growth of nerve cells and their connections depend on receiving particular types of stimulation at specific times during their development. For example, Breedlove (1997) has shown, in rats, that the experience of copulating alters the size of nerve cells in the bulbocavernosus nucleus of the spinal cord, which innervates muscles that are active during the act of copulation. This means that stimulation of activity in these nerve cells alters their structure. Another example is learning and memory. In neuroscientific terms, learning and memory are manifestations of changes in nerve cell connections. In other words, learning requires the expression of genes within nerve cells and, of course, it requires changed behavior as the result of experience. This is a classic example of the interaction between gene expression and environmental stimulation. We are just beginning to understand some of the complexities involved.

In some cases behavioral and cellular changes dependent on experience have been linked. We will summarize an example that one of the authors (Rogers) has worked on for many years. It began with the discovery of hemispheric specialization in the domestic chick (Rogers and Anson 1979), which formed part of the evidence against an earlier assumption that hemispheric specialization was a characteristic unique to humans (see Rogers and Andrew 2002, for subsequent evidence demonstrating this characteristic in other vertebrates). Hemispheric specialization has been of particular interest in the field of sex differences in humans because there have been many suggestions that the degree of specialization (or lateralization) differs in men and women. Most have claimed that men have more strongly lateralized brains than women (Levy 1977) but there have also been suggestions of the converse (Buffery 1981). Whether female brains are considered to be more or less lateralized than male brains seems to depend on exactly what function or behavior is being investigated. It is perhaps better to say that the brains of women and men are differently lateralized but, at the same time, recognize that there is a great deal of similarity rather than difference. Marcel Kinsbourne (1980), a leading researcher in the field of hemispheric specialization, recognized that differences between women and men tend to be exaggerated: "Under pressure from the gathering momentum of feminism, and perhaps as backlash to it, many investigators seem determined to discover that men and women 'really' are different. It seems that if sex differences (e.g., in lateralization) do not exist, then they have to be invented" (p. 241). Others agree that there is little consistent evidence for the existence of sex differences in hemispheric specialization in humans (e.g., Hellige 1993). Nevertheless, the belief has held a central position in debates about sex differences, and it continues to hold this position today.

In the chick, and now known to be so in other vertebrates, the left hemisphere is specialized for controlling responses that require consideration before they are executed, as in feeding or manipulating objects, and the right hemisphere is specialized for rapid responses, as used in attack and responding to novel stimuli. Later investigations showed that the lateralization of these responses is matched by lateralized organization of one of the main visual pathways in the chick (summarized in Rogers and Andrew 2002). Now it would be an easy step to assume that the lateralization is programmed in the chick's genome. Research shows otherwise. The hemispheric specializations for control of feeding and attack responses develop as a result of exposing the embryo, while still in the egg, to light during a critical period just before hatching (Rogers 1982). Just two hours of light experience is sufficient.

Chicks hatched from eggs incubated in the dark have no hemispheric specialization for control of these responses and no lateralization of their visual pathways. The reason light experience has this effect is that just before hatching, when the visual system of the chick is beginning to function, the chick is oriented inside its egg in a way that turns its left eye against its body and leaves its right eye next to the shell and able to be stimulated by light that passes through the shell.

Reversing the eye receiving the light experience causes reversal of the hemispheric specialization and of the lateralization of the visual pathways. Hence these aspects of brain structure and function are malleable and develop in response to experience. Genetic expression is also involved, however, since genes determine the orientation of the chick embryo in the egg. The sex hormones also have a role in the differentiation processes involved. High levels of estrogen and testosterone in the embryo prevent light from affecting the lateralization. This model system, therefore, illustrates the complex interaction among gene expression, hormones, and experience, and shows us that we should not be thinking that less complex events are involved in the development of sex differences in humans.

HORMONES AND FLEXIBILITY

One of the important functions of the brain is to control the amount of sex hormones secreted into the blood. Most of the time it keeps the levels within a stable range, although this varies with time of day and, especially in women, with time of month. In addition, the brain sometimes allows the levels of sex hormones in the blood to change fairly markedly, as in the case of stress, when testosterone levels in men decrease or the monthly cycle of estrogen and progesterone is affected in women. In other words, experience has a marked effect on the brain's control of hormone levels. One mechanism by which it does this is by altering the levels of luteinizing hormone and follicle-stimulating hormone released by the pituitary gland. In this way the levels of sex hormones in the blood can change in response to changes in the external environment. Higher centers in the cortex of the brain process information received from the outside world and, by way of nerve cell connections to the preoptic region in the hypothalamus, these higher processes affect the control of the hormones. Even how we behave or what we think about can affect the levels of our sex hormones.

Despite the fact that much of the thinking and writing about the biology of sex differences implies that sex hormones are a biological entity causing behavior, and not vice versa, a great deal of evidence shows that the levels of the various sex hormones are affected by the individual's behavior. Sex hormones are therefore an aspect of sex differences but they lie at the interface of nature and nurture. This two-way direction of influence is important to remember because, erroneously, the chain of causation is seen to be from genes to hormones to behavior, as if the hormones were merely intermediaries by which the genes exert their effects. Sex hormones cannot alter the nature of the genes but they can determine which ones will be expressed, and thus which proteins will be made by those cells in which they act.

In Chapter 3 we discussed the interaction of hormones and environmental stimulation in the developing rat. As we said there, Moore and her colleagues (1979, 1992) have shown how the effects of testosterone on development of sex differences in rat pups are mediated by changing the behavior of the mother rat, rather than, as first thought, by acting directly on the developing brain. Study of such a model system and the results obtained should have made such an impact on our thinking about sex differences that those adhering to the hypothesis of the sex hormone acting directly on the brain to channel it along a male versus female developmental path would be silenced forever. That has not been the case, largely because scientists, like others, can and do turn a blind eye to evidence that conflicts with their preferred theories—that is, until they are prevented from doing so any longer.

We will give another example showing the interaction of genes, sex hormones, and experience in determining the development of a part of the brain called the corpus callosum. In a postmortem human brain it can be seen as a large white mass between the hemispheres, revealed by pulling each hemisphere sideways a little from the midline or by slicing the brain through the middle. The corpus callosum plays an important role in hemispheric specialization because it allows a region of the cortex in one hemisphere to suppress the functioning of its equivalent region in the other hemisphere (Denenberg 1981). The corpus callosum, therefore, has a quite central position in debates about sex differences in hemispheric specialization, and a number of researchers (e.g., Allen et al. 1991, de Lacoste et al. 1986, Witelson 1989) have attempted to show sex differences in the size of parts of the corpus callosum: note that we say "attempted" because the data collected so far have not been convincing (see Bishop and Wahlsten 1997).

An extensive and important series of experiments conducted by Victor Denenberg and colleagues (1989) has shown that the size of the corpus callosum in the rat brain depends on the interaction of genes, hormones, and early experience. They found that the cross-sectional area of the corpus callosum in the rat is greater in males than in females. The development of this sex difference involves the interaction of sex hormones with experience in the first few weeks after birth (Denenberg et al. 1991). The size of the corpus callosum in females increases if they are castrated just after birth (thus lowering the amount of estrogen in the bloodstream). It can also be increased in males by subjecting them to a particular form of experience, referred to as "handling," in their early life. It involves taking each pup away from its mother for about three minutes each day, when it is placed alone in a small container before returning it to its mother. If the pups are handled for the first twenty days of life, the corpus callosum enlarges and brain lateralization is also affected. In fact, when they are adults, the handled males show a greater degree of lateralization than those that have not been handled. In females this handling has no effect unless they are also treated with the hormone testosterone (Fitch and Denenberg 1998). Male pups, of course, secrete sufficient testosterone themselves, which explains why handling alone is sufficient to increase the size of the corpus callosum of the male.

From these results we can conclude that the experience of handling interacts with the presence of the sex hormones to determine the size of the corpus callosum. It is not known exactly what aspect of the handling experience interacts with testosterone to bring about the increase in the size of the corpus callosum. It could be the effect of brief isolation experienced by the pups themselves or it could be a change in the mother's response to the pups when they are returned to her, since she licks them more after they have been returned. We are reminded again of Moore's research showing the importance of anogenital licking by the mother on sexual differentiation in rats (Chapter 3). It is possible, therefore, that rat pups raised normally by their mothers (without handling or hormone treatment) develop the sex difference in the size of the corpus callosum because their mothers lick the male pups more than they do the female pups. This differential could occur very soon after the pups' birth since nerve connections can grow rapidly at this stage and other changes (such as myelination of the nerve axons) are also more rapid.

Regardless of exactly how handling affects the pups, these results demonstrate a very important principle of development. They show that the development of the brain (in this case, the corpus callosum) depends on the combined effects of experience in early life and the particular hormonal condition of the pup at that time. Sex differences in the size of the corpus callosum emerge because males and females have different hormones in the bloodstream during the early period after birth, but the effects of experience interact with this hormonal influence.

The age of the subjects at the time they are exposed to handling and sex hormones is important for development of the corpus callosum. Handling has no effect on the size of the corpus callosum after twenty-one days of life but estrogen levels are important around day 25 (i.e., after the main time when testosterone has its effect; Fitch and Denenberg 1998). Even in adults the size of the corpus callosum is not entirely static. Researchers in Denenberg's laboratory have also found that a measured sex difference in the size of the corpus callosum of humans varies with age. These age-dependent changes might also result from interactions between hormonal condition and experience. At no time in an individual's life, it seems, is the corpus callosum, or possibly many other regions of the brain, unresponsive to the effects of hormones and experience. The brain retains a degree of plasticity throughout life, although it is most malleable during early development. As the retention of plasticity in adulthood has been recognized for only the last two or so decades, there is much more research to be done on it, especially on the human brain.

All too readily, many researchers opt for biological determinist explanations, even when they have not looked at the developmental processes. For example, de Lacoste and colleagues (1986) assumed that their reported sex difference in the size of the corpus callosum was caused by the action of sex hormones. To quote from their paper: "We believe that our data provide further indirect evidence that the gonadal steroids and/or genetic sex play a role in the development of neural structures, linked with 'cognitive functions', in the human brain" (p. 95). They did not provide any data that even hinted at the causation of the difference.

In an earlier paper, de Lacoste-Utamsing and Holloway (1982) reported that the area of the splenium (the back part of the corpus callosum) was somewhat larger in women than in men, a result that was later also found by Laura Allen and colleagues (1991). The sex difference in size was interpreted as reflecting a lesser degree of lateralization for processing spatial information in women than in men, on the assumption that women might have more connections between the hemispheres in this region. Although no evidence has been provided to support this speculation, many writers in the scientific and popular media have persisted with this idea and have extended it to other behavior that has no proven relationship to brain lateralization or the corpus callosum. This is not to say that the corpus callosum has no importance in brain lateralization, for indeed it is important, but rather that the particular functions and the sex differences attributed to it may not be correct.

Witelson (1989) reported that all the regions of the corpus callosum, apart from the splenium, are larger in men who do not have a strong hand preference than in men who are consistently right-handed. This hand-preference effect was not found in the brains of the women studied. The validity of these results is, however, in some doubt because other studies have failed to replicate them (Aboitiz et al. 1992). Also, studies that have used the fMRI technique to measure the areas have failed to find any significant effects of handedness or sex on the size of different regions of the corpus callosum (Bell and Variend 1985). Considering all of the studies that have measured the size of the human corpus callosum together, we can only conclude that any early indications of the existence of a sex difference in the size of the corpus callosum have not been substantiated, as reported by Bishop and Wahlsten

(1997). The latter researchers carried out a comprehensive reassessment of all of the data on the size of the corpus callosum collected up until 1997 and concluded that there is no consistent meaningful sex difference in its size. In the majority of studies the sample sizes have been too small to be representative of a population. Although the corpus callosum in its entirety may be larger in men than in women, on average, men also have a proportionally larger brain size and body size than women. When this is taken into account by expressing the size of the corpus callosum as a ratio of body weight, any potential sex difference disappears.

Here we are reminded of the research in the 1960s on the effects of testosterone in early life on the developing rat brain, and we want to return to it briefly because it has such a pivotal place in thinking even today. As Beach (1971) pointed out then, much of the interpretation of the research was distorted, seemingly to confirm the original hypothesis of Phoenix and colleagues (1959), which claimed that development of a male brain depends on an organizing effect of testosterone during early life and that without this exposure, a female develops irrespective of genetic sex. This has become known as the organizational effect of testosterone. In Chapter 3 we discussed how a number of studies claiming to provide data in support of this hypothesis failed to take into account the fact that the effective action of testosterone in this respect is on the behavior of the mother rat, rather than directly on the pup's brain. We recognize that an effect of testosterone on the rat pup's brain may also be important but it cannot be an overwhelming effect because females behave like males after the mother has treated them like males in early life. Researchers of the 1960s embraced the organizational hypothesis so enthusiastically that it was extrapolated directly from rats to humans, becoming the explanation not only of sex differences but also of homosexuality, lesbianism, transsexuality, and transvestism (discussed in Rogers 2001). The organization hypothesis became orthodoxy and still persists as such in many quarters of biology and psychology, as well as making frequent appearances in the media, although some (male) researchers have just begun to see its pitfalls (e.g., Breedlove 1999) two decades after critiques by women scientists (Bleier 1984, Rogers 1981, 1988). The brain may be organized in early life but it is not testosterone alone that does it.

CONSIDERING INFLUENCES

The genes are not the code of life and the hormonal mix is not a blueprint for development, as some have claimed. The effects of genes and hormones should not be seen in isolation from external inputs to the developing system. Similarly, inputs from the environment (that is, experience) should not be considered in isolation from the genetic or hormonal factors. No single factor has primacy in directing development. It follows, therefore, that genuine understanding of the processes of sexual and gender development will come only by looking at the interactions among all of these factors. This principle would apply no matter what aspects of sex or gender difference a researcher might choose to focus on, be these molecular or cellular aspects of the brain itself or the manifestation of brain function as behavior. It is also important to look not just at differences that might take place at one stage of a person's life, but also to look at the development of these differences.

The premise that one biological fact, such as expression of a gene or set of genes, weight of a brain, number of receptors, or absence or presence of one specific hormone has broad explanatory power and gives some insight into the behavior of an individual is false. In any field, monocausal explanations of scholarship rarely stand up under scrutiny. The same is usually true of explanations that place greater emphasis on one or some of the causal factors at the expense of others, and on failing to consider the interactions between these factors.

For the study and explanation of human behavior it is logically and scientifically impossible for one small facet to explain all when there are a multitude of variables that make up the human environment and human behavior. That is, of course, the challenge and the difficulty in studying human behavior. In terms of experimentation, human behavior is less accessible than is the equivalent in animal studies; when a theory seemingly makes the complexities disappear, there may be a sense of relief that manageable answers to our questions are finally at hand. However, we believe in the need for skepticism and rigorous science instead of adopting unsubstantiated conclusions and prematurely maneuvering results to accommodate prevailing theories or influence social attitudes. Wider recognition of this need will change our thinking in science as well as in society.

PART TWO

In Realities

Either nature has made a great difference between man and woman, or . . . the civilization which has hitherto taken place in the world has been very partial.

M. Wollstonecraft, 1975

5

Gendered World

In the previous chapters we have spoken about sex and gender differences in the context of biology and society but we have not yet discussed these concepts with reference to social practices and intellectual traditions.

POWER WITHOUT

We are only too aware that social life in human societies is divided according to gender and that few, if any, can claim to have moved entirely away from gendered practices. For most sociobiologists and evolutionary psychologists, this is proof that the division between male and female is immutable in all of its forms of expression. For them, the ubiquity and persistence over time of gender divisions is sufficient evidence that biology has its imprint on fundamental differences between the sexes. But historically, similar claims have been made about slavery, which is almost as old as human life itself (Winks 1972). Yet the world community now agrees that slavery is not a biologically determined category and that it is not an immutable state of being. It was abolished in the Western world less than two centuries ago, although it still lingers on in some other parts of the world (Archer 1988).

In studies on slavery, and even on women, two words used regularly are *exploitation* and *power*, with one important conceptual addition for women, and this is *sexuality*. It is a third dimension in which exploitation and power are played out in more complex, at times interdependent, and often ambivalent ways. A slave could be but a slave, a person stripped of status and dignity, subjected to orders and forced to obey in a litany of humiliating deeds spurred by mutual hatred and contempt or, at a distance, maintained by indifference or thoughtlessness. Similar power differentials, when applied between men and women, may be tempered or heightened by sexual relations. Even in the worst bonds—and then against the will of the actors—desire, dependence, and passion may be mixed into the brew of contempt, subjugation, and despair. Love is rarely a component of either slavery or of exploitative sexual relationships but power is its axiom.

Since the publication of Michel Foucault's *A History of Sexuality* (1978) we know that freedom from oppression, and, much more deep-seated, from repression, may not be just a matter of lifting a few prohibitions. Sexuality, he argued, does not exist beyond power relations and is not controlled by centralized power. It is actually produced by those power relations that both repress and saturate it. Foucault's famous dictum is "Power is everywhere; not because it embraces everything but because it comes from everywhere" (1978, p. 93). "Everywhere," he would agree, is male defined.

The first thinkers of the modern era to concern themselves with power as a basis for inequality were those of the enlightenment in the eighteenth century, culminating in 1789 in the French Revolution with its battle cry for liberty, equality, and fraternity. However, with few exceptions, even the most enlightened thinkers were generally unable to conceive of women and men as biologically and socially equal. Instead, they thought about how to create harmony within a household and in marriage. Rousseau, one of the giants of enlightened thought, outlined his conception of the perfect female in *Émile*, first published in 1762. He wrote: "What is most wanted in a woman is gentleness: formed to obey a creature so imperfect as man, a creature often vicious and always faulty, she should early learn to submit to injustice and to suffer the wrongs inflicted on her by her husband without complaint" (p. 328).

Although Mary Wollstonecraft, his contemporary, was outraged by these sentiments and dismissed them (the above passage especially) as nonsense (Wollstonecraft 1975), the belief continued to be a kind of ideal throughout the nineteenth century for women and men alike, and so continued to legitimize and deliver potentially unlimited power to the male head of household.

Only in some literature and again since the post-1970s were such male ideas of wifely obedience seriously tested in art and life. There is a forceful passage in Irini Spanidou's novel *God's Snake* (1986) that plays out a possible scenario of Rousseau's dreamings in *Émile*. The girl/child narrator recounts the sad ending of Kyroula, the dog, who was "disciplined" by her father with heavy beatings:

> As the beatings did not succeed in making Kyroula faithful or ferocious, my father locked her in the woodshed, a lean-to without a window four meters long and one meter wide, and ordered Manolis to feed her bread and water once a day. He wanted to awaken her savage instincts. . . . In the beginning she was let out every two or three days. Blinded by the sudden light, her legs staggering from weakness, she crouched toward my father, peeing along the way—incontinent, docile, limply wagging her tail. When he shut her in again, she whimpered, resisting, legs rigid, pushing helplessly on the ground, so he had to drag her. In the end, she was not let out at all. Her wails, low, long yowls fading in a despair of silence then starting again like moans of anguished, unbearable pain, filled the house night and day. Everyone pretended not to hear. . . . When one day no more crying came from the shed, my fear, like the silence, would not end. I paced in the house terrified.
>
> [Then the feverish starving dog was carried to the father and put at his feet, whereupon the mother said to the father]:
>
> "You're poisoned with your own venom and want to see everyone else poisoned and destroyed, crawling at your feet," my mother said. "Power! All you want is power. You don't have it, Stephane. You can kill, you can destroy. You killed me. But you don't have power."
>
> "Who has power?"
>
> "Love has power."
>
> "Such as yours?"

> "I have no love. I have no self-respect anymore."
> "Did you ever?"
> "Before you trampled my soul."
> "Words."
> "Kyroula? Is Kyroula words?" [pp. 155–157]

Could one put the absolute demand for power more succinctly? Kyroula is not just a name; she is a dog dying of the brutality of the patriarch and, worse, this is possible because no one dares to bring her food or free her from the shed. The mother is everything that Rousseau demanded of Émile: she does not interfere in her husband's affairs but her compliance is poisoned. For the reader, the events are morally outrageous. The home is devoid of harmony—which the woman's subservience was meant to achieve, according to Rousseau (Weiss 1990)—and, instead, it has become a killing field. He destroys his wife's soul, terrifies his daughter, and slowly kills his dog for the sake of a specific kind of obedience, not just for crouching and subservience but for an obedience that must not show resistance even in extremes of fear, thus demanding an abandonment of emotions altogether. Yet his wife has the strength to deprive him of victory in the face of his senseless act of rage. The author has presented a powerful case demonstrating why not all is right with Rousseau's ideals. Impressively, she has managed to portray an image of a household in which everyone but the patriarch is homeless. The wife and the daughter are homeless because there is no refuge, no respite, no safety, and no love.

IDEAS—THE PAST

Homelessness of women, implicitly linked to power, is a theme running through much of the feminist academic writings from the 1970s to the present. Feminists of the modern era have searched for an intellectual and an emotional home within the intellectual traditions that women have inherited. Okin (1980) saw this not as "an arcane academic pursuit but an important means of comprehending and laying bare the assumptions behind deeply rooted modes of thought that continue to affect people's lives in major ways" (1980, p. 3). Western thought has been examined from Plato to Nietzsche. In the last decades of the twentieth century, specific attention has been given to Plato (Bluestone 1987, Saxonhouse 1976, 1984), Machiavelli (Pitkin 1984), Hobbes (Di Stefano 1983), Locke (Clarke 1979), J. S. Mill, Foucault, and many others (O'Brien 1989). Rousseau underwent almost a decade of considerable attention and feminist criticism (Schwartz 1984, Weiss 1990). There has also been renewed feminist interest in Hannah Arendt's work, this time not so much perpetuating Adrienne Rich's early criticisms but rather as an engagement with Arendt and as worthwhile attempts to make her part of a new feminist political theory canon (Elshtain 1986, Honig 1992, Kaplan and Kessler 1989).

Feminists have identified a deep structure of masculinity in the history of ideas (Di Stefano 1991). Writers as different as Hobbes and Marx are shown to be ultra-conflictual while even the more egalitarian-minded ideas of John Stuart Mill have bent to the influence of a culture of masculinity in his notion of a plastic, self-creating identity. Benhabib (1989) complained that the sexual division of labor in the family, meaning reproductive rights, child rearing, and socialization, are ignored, romantically trivialized, or linguistically whitewashed. Mainstream theory, feminists argued, had failed to theorize the body (Eisenstein 1988), and hence was incapable of dealing with specific questions such as rape in marriage, or marital "slavery" (Shanley 1981).

Much of gender theory has been devoted to exploring the legal and social basis of female/male sexual and marital relationships. Early on, this led to innovative analyses of the concept of consent in contract theory as applied in male-female relationships and marriage (Pateman 1988). Pateman suggested that consent theorists had avoided the difficult question of how consent to rule occurs and who consents. She proposed that unless withdrawal of consent is possible, consent is meaningless. In sexual relations, the link to power and subjugation is clear. Understanding this nexus inevitably leads to a substantial departure from traditional thought on the nature of sexual relationships between women and men, and on power. Hobbes had in fact argued that consent, *even when induced by fear of retribution*, is equivalent to free consent. Hence, in traditional philosophy (also in modern legal structures) the relationship between men and women resembles an ideal of female subordination modeled on a master–slave relationship more closely than it does a mere contractual inequality. The perception is clearly consistent with that of Rousseau's *Émile* (see above). Another subplot to the feminist debates on contract has come from radical lesbian groups, arguing that the social contract compulsorily assumes heterosexuality and heterosociality (Wittig 1989).

In other words, the Western intellectual tradition does not actually speak about the human condition as if it concerned all humans; it is gender specific in the way in which it has developed concepts, including gendered dichotomies. For instance, Descartes' distinction between reason and non-reason on the basis of intellect/feeling and body/mind linked the notion of reason to gender differences (Lloyd 1984, 1991): intellect and mind became "male" and body and feeling "female." From ancient times onward, so Pateman (1988) argues, political life "has been conceptualized in opposition to the mundane world of necessity, the body, the sexual passions and birth" (p. 115), and hence *in opposition* to women, who were regarded as part of that mundane world. Authorities of the stature of Rousseau or Hegel expressed cognizance of women's ability for love and nurturing, but both authors derived from this insight the conclusion that women, being "naturally" different from men (Weiss 1987), had to be excluded from the public realm of citizenship *because* they are caretakers "of affectivity, desire, and the body."

Reason, so it was argued, is based on impartiality, and impartiality is central to an enlightened morality. Emotions and desires are *ab ovo* incapable of achieving impartiality; since women are the carriers of those specific emotions they must not, and, by their nature cannot, enter public life. According to enlightenment ideas, and also to modern ethics, affect and desire have to be excluded if moral reason can be achieved. Young (1986) argues that "this ideal of impartial and universal reason generates an oppressive opposition between reason and desire or affectivity" (p. 382) and thereby also derives one of the central splits from these dichotomies: the distinction between the private and the public, the private realm being identified with women and the public with men. This gendered public/private distinction has been upheld by most thinkers and is still being vigorously defended by some (such as Jürgen Habermas; see Fraser 1991) and, of course, it was one of the main theoretical starting points of debate in the second wave of women's movements in the 1970s.

Even (male) thinkers who conceived of the notion of equality in the context of citizenship continued to exclude women from such equality, either consciously for reasons of social utility and political expediency, or subconsciously as an omission and an undetected paradox on the part of the writer. Rousseau has certainly been credited with the former (Weiss 1987) and John Stuart Mill and even Hobbes with the latter. Pateman (1991) identified a paradox in Hobbes' thought by showing that, although he acknowledged the equality of men and women in the state of nature, he viewed women's role in the family as domestic slaves. Having taken this leap from nature to home he was therefore able to preclude women from any social contract altogether. In Tocqueville's model of democracy a central paradox is the equality attributed to women but the limited role and activity allowed them (Winthrop 1986).

It also emerges clearly in this intellectual tradition that certain actions and characteristics were thought of as masculine or feminine. War and peace have often been assigned a gendered position. War is considered a purely masculine idea, peace attributed largely to femininity. Especially from Hobbes onward, coalescing at first in Machiavelli's ideas expressed in the Renaissance, aggression and war occupied a considerable part of the Western intellectual tradition. Men have developed models of the so-called good war or just war. Hegel's nineteenth-century view of the inevitability and healthiness of war is not outmoded today. Ideas were thus carried into the public arena (from which women were excluded) and from there to the battlefield. The predilection for conflict, feminists have argued, owes much to the pattern of masculine self-development within patriarchy (Fuchs 1983).

IDEAS—THE PRESENT

Beginning in the mid-1970s, similar views on aggression and good wars have been revived via an entirely different intellectual route and have made their appearance in the writings of sociobiology. Sociobiology regards aggression as genetically innate in males and therefore as "natural" and defensible. Aggressive behavior is said to be a male characteristic and, therefore, controlled by genes on the Y chromosome. In 1981, Lumsden and Wilson suggested that genes involved in social behavior be called "culturgens." War and the means of making war were allegedly based on such culturgens. Pinker and other evolutionary psychologists, including Tooby and Cosmides, have linked aggression, hunting, and war. Hunting, they argue, has been a major force in the evolution of the human mind (see Barkow et al. 1992), an idea borrowed from a long history in anthropology. In their equation, necessity for hunting as a male activity (due to the male's alleged higher, hormone-induced aggression) also means men "naturally" have to fight each other. One notes that it is an aggression not just directed toward war but explicitly against women. This becomes clear in the "rational" manner in which men's rape of women is "explained" (Thornhill and Palmer 2000). Violence, so we are told, may have been a route to sexual success for our distant ancestors, especially in times of turmoil. Matt Ridley (1994) regarded this as an explanation of why rape is still common in war to this day. Pinker spun out this tale even further by claiming that men banded together to fight *because* a conquest in tribal war allowed the victors to rape the women and so pass on their genes.

The word *power* has gone from the vocabulary and *sexuality* is disembodied entirely because in this contemporary theoretical world the authors think, mistakenly, that they are dealing only with essential matters: the passing on of genes. Sexuality and power slip in at the backdoor, however, through pondering the alleged needs and lifestyles of cave men! We had always thought that we had progressed so much in the last 100,000 years or so! Apparently, this is not so because our brain today is said to be the same as human brains one or two million years ago. Not that we could ever disprove or confirm any of these claims, let alone assess their relevance to human evolution, but, importantly, it is a tale made in the present for the present and told in a morally devoid way. The parallels to economics are all too obvious. Sociobiology and evolutionary psychology, as in economic rationalism, use a beguiling simplicity and matter-of-factness of language. The natural world is neither moral nor immoral; hence it is just a matter of describing it. Suffice it to say that their explanations of conflict, war, and aggression are, ultimately, placed beyond criticism. This is achieved by describing these characteristics as outcomes of a genetically determined male "nature" (Dawkins 1976, Wilson 1975; for a critique see Rose 1984).

The assumption of the innateness of aggression in men is problematic, to say the least. Agnes Heller called the term *aggression* a bag into which anything could be placed. Aggression was rarely ever defined and served ideological rather than scientific purposes (Heller 1977), an observation that equally holds true today. Whatever its attributes, even in ill-defined ways, the assumption of some kind of genetic basis for aggression has vast social implications, not least of which actually appears to be an implied plea for its social acceptance.

The explanations for violence, aggression, and rape have several major flaws, only two of which we want to address here. First, when thinkers of the past spoke about conflict and even about "good" wars, as did Hegel, there was still an assumed imperative of morality underpinning their theories. The moral base has gone because some current thoughts on violence have shifted from human history, laws, and culture to general biology and nature. Hence the ballast of thousands of years of thought in the social domain has been left behind. This has happened only once in recorded history, during the period of fascism, when biology indeed became destiny (Bridenthal et al. 1984).

Some theorists working in evolutionary psychology may need reminding of Aristotle. He held that the distinctiveness of human animals lay in their power of reasoned speech, and that this faculty facilitated the articulation of moral values and helped constitute the human as a moral being, admittedly also an essentialist position for which, however, defensible reasons have been found. The purpose of a human life was to fulfill its ultimate object or aim *(telos)*. Since intellect was thought to constitute human nature, so the development of intellect was the *telos* of fulfilling human life and living morally (Aristotle, *Politics* 1.2, 1253a 10–20). There is no evidence of intellect in the game of genes. Intellect, for Aristotle, endows humans with power for moral action and agency while, for evolutionary psychologists, the intellect is a mere byproduct of the action and purpose of genes.

Second, if we descend from Aristotle to the cave men just for a moment, there is little plausibility in the belief that rape and passing on genes were intricately linked events. Many women have been raped through the ages but few actually conceive: (1) because usually women are not reproductively receptive for more than about 2 to 5 days per month and the rapist would have to be lucky to catch the right moment in the monthly fertility calendar; (2) women who are being raped are so fearful and have such strong physiological responses that the body can usually prevent successful implantation of the ova; (3) there is a naïve belief in some of the writing that the women who had been unlucky enough to conceive as an outcome of rape would actually bear a child so conceived and raise it. Abortions were not just inventions of the modern era. Moreover, infanticide was a practice dictated by need and, sometimes perhaps, by hate. Women could not be relied upon to be the vessel to give new life to the enemy's genes and it would seem to us to be rather ludicrous to suggest that it would have been a reliable tactic, if passing on genes was a goal. It further appears that evolutionary psychologists do not believe that women can take actions of their own. And finally (4), evolutionary psychologists and sociobiologists tend to overlook a somewhat crucial point: human societies of the distant past appear not always to have known that the male of the species had anything to do with reproduction. In many cultures, women were seen as the givers of life but the role of the male or of sperm was not necessarily understood (Frazer 1974).

The further back in time we go, the less likely it would have been that sex and reproduction were understood as being related. Even in more recent human societies, the idea of impregnating the enemy may not have worked as well in practice. The above scenarios, speculative as they are, at least should invite substantial skepticism about the hypothesized genetic selection of Stone-Age minds!

This is a far cry from the power of reason that Locke advocated and described as the true state of nature. He said: "Men living together according to reason, without a common superior on earth, with the authority to judge between them is properly the state of nature" (cited in Russell 1945, p. 625). Even Hobbes, who had no illusions about human "nature," saw that without reason and without government man would be thrown back into a "natural state," leading him to conclude, in his famous words, that in such a state of nature "the life of man [would be] solitary, poor, nasty, brutish and short" (Hobbes 1955). It appears that evolutionary psychologists want us to believe that we are in our minds still in such "a state of nature."

Remarkably, one of the strongest inconsistencies in the gender schema of evolutionary psychology—and previous intellectual traditions—can be found in the gendered allocation of emotions. We have said before that throughout the Western intellectual tradition, emotions (the affective domain) were considered feminine traits while reason and objectivity were considered masculine. Yet in the very same breath scientists and philosophers alike have argued that aggression is a distinctly male domain. For the past twenty-five years we have seen publications promoting the idea that aggression belongs to the male of the species and that there was some evolutionary advantage for the male to be so inclined. Few would be able to argue against the idea that aggression belongs largely to the emotional domain (even if planning aspects may appear rational). Hence, in the way the human characteristics are parceled out, it appears that men do enter the affective domain that philosophers had thought belonged to women. Presumably, aggression has been identified with men to justify war and power struggles as a *genetic* quality. One suspects, however, that a justification for power and sexual dominion over women might well form a centerpiece of these hypotheses.

On the question of war, even some radical feminists have accepted that women can represent a different voice (i.e., one for peace) but this was largely a result of a new feminist consciousness that perceived human liberation and peace as interlinked (Elshtain 1981). Patty Walton (1982) wrote:

> Women have not fought in wars because of our material circumstances and not because we are innately more moral than men or because of any biological limitation on our part. The work of women supports both a society's war and its peace activities. And our support has always derived from our particular socialisation as women. In fact, the socialization of women and men complements the needs of the culture in which we live. It is necessary to recognize this because we need to change these material relationships and not just the sex of our world problem-makers. Men are not more innately aggressive than women are passive. We have cultures of war, so we can have cultures of peace. [p. 43]

Sadly, so feminists discovered, there were very few male thinkers whose ideas had elements of such broader humanist or feminist visions of humanity. One exception is Mill's *The Subjugation of Women* (1869) for his notion of friendship, albeit inadequate, which made at least some attempt at raising issues of equal opportunity for women (Shanley 1981). Mill, it is argued, wished the family to be the modern version of the polis, forming the basis for justice in human relationships (Urbinati 1991). Mueller (1988) suggested in *The Politics of the Gender Gap* that women's history of less active participation in public life is related to their traditional lack of access to education and lower income. However, propositions that hint at a deficiency model of women perpetuate notions of women's inferiority.

With hindsight, therefore, it is easy to see that contemporary women find little solace by looking back and reading the works of even the greatest thinkers in history. Instead they have found a hostile place where women—half of humanity—simply do not exist or can be used and exploited. If we now interrupt at this point and ask the reader to remember that "new" theories on the alleged genetic basis of aggression and caring have been resurrected in sociobiological and evolutionary theories, it takes little imagination to slot those "new" ideas into a long, distinguished, but strongly gender-biased history of Western intellectual traditions. It is worth emphasizing that, whereas the more persistent oppression of women has received support from molecular genetics and now evolutionary psychology, the liberatory movements of the modern era have generally had no support from biology. Indeed, despite the revolutionary new findings in biology in the last decades, it seems to us that, overall, the field has reconfirmed, and in some cases resurrected, biased and conservative notions that we thought had long been laid to rest.

In this climate, there is no refuge for women to draw on past authority, and no home for a moral outrage, seemingly, in the pseudoscientific world of gene supremacy. There was no home for Kyroula because there was no one with agency to intervene. Marcuse argued that the liberatory impulse of the social process lies in the dynamic of negativity (that is, in the negation of existing norms) and in making the absent present because the greater part of the truth is precisely in that which is absent (Marcuse 1968). The women's movements, as well as other new social movements, provided this "dynamics of negativity" precisely in order to create agency (Cornell and Thurschwell 1987).

POWER WITHIN

Despite the gender-bias in the history of ideas, we have evidence there have always been some short periods in history when social practice was more favorable to women and rarely, but importantly, these translated into new ideas. These were the seventeenth-century so-called mid-century revolts and the French Revolution. Between 1647 and 1649 the Levellers had drafted a type of constitution that they called *Agreement of the People* (Walzer 1974). This *Agreement*, one hundred years later, formed the basis for the American *Declaration of Independence* in 1776. The American *Declaration of Independence* in turn was the basis for the feminist *Declaration of Sentiments* in the United States in 1848. Indeed, the latter was closely modeled on the former and so was the ensuing feminist *Declaration of Rights for Women* of 1876, which demanded justice and equality, and most notably "that all the civil and political rights that belong to citizens of the United States be guaranteed to us and our daughters forever." It was Hannah Arendt's (1972) contention that, among other things, the American *Declaration of Independence* was a profound political act because it constituted a new political community. *Mutatis mutandis* one might apply this assessment to the landmark declarations and manifestos issued by women; in a sense, their manifestos, too, created new political communities (Oakley 1972).

By the late 1960s the claims had become far more radical, and the demands and arguments more all embracing than they were in earlier declarations. Although these new women's movements were largely political, they were also concerned with society, nature, and gender roles in the broadest sense (Kaplan 1992, 1996). Many took a forceful stance *against* biological explanations for their behavior. Generally, they refused to accept the notion of biological status as a yardstick for their social, intellectual, and political roles. These women's movements across the Western world have not accidentally been called the longest revolution (Mitchell 1966, Pasquinelli 1984).

The other part of the English-speaking feminist tradition begins with a tenuous link to the French Revolution and in the eighteenth century, to the Enlightenment via Mary Wollstonecraft, who belonged to the small circle of radicals supporting the Revolution. Her criticisms of Rousseau's misogyny and sex-role assignments are a starting point for a long line of feminist theory extending until this day.

Natasha Walter took up these threads when she reminded her readers of this traditional saying that feminism had "overpersonalised the political and overpoliticised the personal" to the detriment of "two great, longstanding goals: political equality and personal freedom." Indeed, regardless of women's personal or political beliefs and the moment in time, political and social equality and personal freedom have been goals of enormous importance for women all over the world. It is particularly pertinent to put this reminder here, as some recent books on sex and gender either have not been very clear or have shown exceptional ignorance about the word and application of *equality* (e.g., Kimura 1999) and so have many polemical writings against affirmative action, or simply against perceived injustices, later termed depreciatively "politically correct" attitudes (Schwartz 2001).

In 1999, Kimura argued polemically against the "ideology" of egalitarianism because, in her view, it was a belief without empirical evidence. She substantiated this by saying that she had grown up with her brothers and they were all different despite the common environment. Kimura, and many others interested in stressing biological differences between men and women, have failed to take into account a 2000-year Western tradition of trying to arrive at and define notions of justice and at least a 200-year intellectual history tackling potential equality as a political and civil right of all human beings. It is not relevant to that tradition whether brothers were different from sisters, or from each other, when they were growing up. Morality does not require empirical evidence. Egalitarianism is a moral position for a polity, a state, and a nation, and an ideal for personal freedom and civic equality. There is nothing in law books that states that we must be twins in order to be given the same legal treatment. For those unfamiliar with the basis of much of Western law, we wish to note that Marcus Tullius Cicero, Roman statesman and famous orator under Caesar, wrote books about the principles of law in the first century BC that have become the blueprint for jurisprudence (Roman Law) apart from Natural Law and Common Law traditions. In his famous *De Legibus*, Cicero wrote:

> No single thing is so like another, so exactly its counterpart, as all of us are to one another. Nay, if bad habits and false beliefs did not twist the weaker minds and turn them in whatever direction they are inclined, no one would be so like all others. And so, however, we define man, a single definition will apply to all. . . . For those creatures who have received the gift of reason from Nature have also received right reason, and therefore they have also received the gift of Law, which is right reason applied to command and prohibition. [1951, Vol. I, x, 29; xii, 33]

The gift of reason, noting here the Aristotelian influence, is thus the unifying sense of humanity; whosoever possesses reason may therefore receive the law as an equal. Reason is the broadest qualifying way of describing the human species. This is the essence of egalitarianism as it was first expounded by jurisprudence and seen as the very foundation upon which a state can administer its realm with any sense of consistency and order. Equality before the law on the grounds of the faculty of reason alone has been one of the oldest and strongest pillars of legal practice for no less than 2000 years. One may forgive our surprise, perhaps, to find that such principles are now treated by some writers as if they were the brainchild of a few contemporary radicals. The great social and political struggle has been to get rulers and thinkers to agree and to admit that reason is also a faculty possessed by those who were not property owners or clergy (as was the case before the French Revolution), who may be working class and property-less. In the nineteenth and twentieth centuries the new questions asked were whether people of different ethnic backgrounds, different color, and even women might not also possess reason—reason enough to be educated and reason enough to vote.

The great *moral* question that many lawmakers, politicians, and philosophers have asked in the last 300 years is whether people should go to jail because they have no money to defend themselves or whether the punishment of jail should be reserved for those who have been proven to be guilty on good impartial evidence. The latter point, if answered in the affirmative, is based on assumptions of egalitarianism: that is, a belief in equality before the law which, in turn, is based on the highly cherished belief since the Enlightenment period (eighteenth century) that impartiality is the basis of justice. If we made a law tomorrow that all men below the height of 5'6" need to go to jail because they are polluting the gene pool, we would have achieved a biological reasoning of some sort, but no judge would possibly claim that this is a proper administration of justice. In other words, the original *idea* of equality is not grounded in biological differences or similarities (unless these mar legal responsibility, as in the case of mental health). Nor, ever since the great French Revolution, should it be argued again that unequal principles ought to apply because someone is black, female, disabled, has a long or a short nose, blond or black hair, and so forth.

Kimura and others seem not to have understood that not all social life can be based on models of biology, because human society has created complex superstructures of abstract ideas, rules, and regulations to manage its social, economic, and political life. Some of these superstructures are geared toward vested interests and maintenance of power.

Others, or at least facets of that society, are humanitarian because they aim toward greater freedom and better treatment of any individual within a given polity. While norms and values are often derived (erroneously) from presumed or unproven biological differences (particularly in the case of women), as we have tried to show in this and previous chapters, egalitarianism is *not* a biological belief but a civic and political one. In fact, it is not so much a belief as it is a longstanding civic agenda and practice.

As the new women's movement of the 1970s well understood, political eguality on its own would not bring liberation. Feminists also had to fight for personal liberty because so much of it is tied up with their roles as citizens and with the public sphere. It may be the freedom to open a bank account, to walk across a public place unmolested, to know that any attack on one's person will have the support of the law to find and punish the offender. It may be that personal freedom is the right to develop one's full potential, both in education and career, or to freely choose a partner or a place of domicile. The United Nations Declaration of Human Rights outlined most of these as a program for a fairer and more humane world in 1948.

Indeed, it is not despite inequalities that egalitarian agendas persist (as Kimura seems to believe) but precisely *because* substantial inequalities exist (Rhodes 1989) that these rules of law and declarations of human rights have come about. They are not designed to make us all equal (intellectually, physically, economically, etc.) but to give everyone access to the same rights and the same justice within a sociopolitical polity/republic. These principles are so very far removed from biology that they should not even be named in the same breath. By implication, Kimura's travesty has been to nearly equate very broad biological principles of nature with management of societies and conducts of the polity. On intellectual and moral grounds, these distinctions had been made rather clear early in Western intellectual traditions, at the latest with Aristotle in the fourth century BC.

Admittedly, there is at work today a vulgar Darwinism resurrecting biological notions of the survival of the fittest. The followers argue to let the weak and old die; to force sterilization on women who are judged to be unfit as mothers (see Silliman and King 1999); to subscribe to rules of aggression and to award sexual freedoms only to men (because they need it); to abort children for reasons of sex, IQ, or even more whimsical reasons (Savulescu 2001); and to subjugate women or people of color because those with power have decided who is fit and who should rule (Ginsburg and Rapp 1995).

One of the greatest assets of nature is the enormous diversity of organisms within it and, among humans, an extraordinary variety within the same species. That is a biological strength but not one from which one should logically derive the need for social (negative) discrimination. The resurgence of arguments for strong innate differences between women and men, which condones or at least implies a host of "accommodations" (to violence, war, aggression, rape) for males, belies our ability to think and change the social world in which we live. Certain propositions, once put, may set new frameworks and then entrap other thoughts within it in such a way that even attempts to break out ultimately function within that framework and use its tools, as Frankfurt School philosopher Theodor Adorno had once argued. Susan Oyama (1985) put it very well when she said that terms like genetic and biological in fact are often effective synonyms for inevitability, unchangeability, and normality. These are essentialist categories that are based on some, but not all, existing philosophical arguments.

GENDER AND "WOMAN"

The development of feminist theory required questioning terms such as *gender* and *woman*, examining the interplay between gender relations and institutional contexts and the way in which gender is knitted into structures of domination (Disch 1991, Love 1991). In such battles about terminology, feminist theory suffered from "normative confusions," to use Nancy Fraser's term (1991) and, more pertinently, from unexpected outcomes. For instance, "simple" questions, such as "Who is a woman?" and "Are there women?" can be dangerous in the hands of male philosophers. Jacques Derrida and Michel Foucault actually made the category "woman" disappear. Just when feminists discovered "woman," Foucault decided to celebrate the disappearance of "man." When women had just begun to fight for social rights and justice, Derrida suggested that "she" did not really exist, that a woman was "a non-identity, non-figure, a simulacrum" (Derrida 1978). Regardless of such complete erasures, questions of identity plagued feminist theory for decades.

The questions were as basic as "Who are we?" and "What do we want?" but they were, in fact, not easy to answer. Some feminists adopted a determinist position. They argued that there is a biological basis for a distinct "feminine" character that is denied existence in a male-dominated society. Their views on biology led them to adopt a standpoint that may best be summed up as a claim for equality with difference. In their political framework, equality of opportunity was not identical with equality and lack of discrimination. Positive discrimination was seen to be necessary in order to take into account the special needs of women as dictated by their biology. Theirs was a position of aiming for slow changes brought about by educating women to overcome some of the biological limitations which, they had been taught, placed them in inferior positions. For example, Lambert (1978) argued that the organization of male and female brains is different and biologically determined; thus any change cannot go beyond this fact. Reform rather than revolution was seen as the only acceptable way to move toward equality. By far the majority of feminists, however, felt that biology had been used far beyond its applicability in the social domain and had, in fact, been the major tool of oppression.

An antifeminist camp also continued to be heard, consisting of women whose views had not been particularly welcomed in the women's movement (Breines et al. 1978), such as Alice Rossi's, for instance. She went about celebrating women's virtues in a way that was appropriately dubbed "biological essentialism" (Sayers 1982, p. 149). These feminine virtues were said to be biologically innate talents, such as the ability for child-rearing and nurturing. Rossi (1965, 1977) argued that these qualities needed to be fully recognized, and only such recognition would actually lead to equality (in the sense of difference as being of equal social value). Incidentally, her arguments of the late 1970s were not by any means original. Fascist ideologues had announced the same principles of "equality" in the late 1920s, even before Hitler came to power (Kaplan and Adams 1990). There still appear to be women who can be persuaded that something as abstract and as dependent on social mores as "virtue" can be tied to biology and, worse, who naively believe that a celebration of feminine virtues would lead to equality. Some defenders of feminine virtues felt they were speaking their own mind by supporting biologist prejudice! At the dawn of a new century, we have women who proudly face the world with the shattering message that a woman might have "a mind of her own" (Campbell 2002). Even animals have "minds of their own" (Rogers 1997) but this is no guarantee of civic liberties either for humans or animals. Men have celebrated feminine virtues for thousands of years without much evidence of women gaining any more rights or freedoms.

Another group of feminists who likewise accepted the biological determinist position were far more pessimistic, or radical (depending on one's viewpoint). In their view, the so-called feminine character had little chance of developing and blossoming in a male-dominated world and could only be cultivated in isolation. They proposed separatism based on biological difference.

Nevertheless, at the onset of the women's movements in the 1960s and 1970s, women expressed a feeling of universal sisterhood, a oneness in experience, and a sense of unity of purpose. "What we want" seemed clear but such certainty was quickly undermined. Black, ethnic, and colonized women began to assert their differences from mainstream feminism. Black women pointed out that mainstream feminism excluded different sets of problems and issues for women of color (hooks 1981, 2000, Smith 1985). Alice Walker (1983) even preferred to drop the term *feminism* in favor of *womanism* in order to signal the ethnocentrism of feminism. Ethnocentrism was then investigated systematically (Spivak 1990, Trinh 1989), and blackness itself became a paradigm of political and feminist resistance against white domination and exploitation (Collins 1990, Mirza 1992). Women from developing or colonized countries also raised objections. A speech by an influential political feminist from Vanuatu may suffice to indicate the diversity of views in feminist ranks:

> Women's liberation or Women's Lib is a European disease to be cured by Europeans. What we are aiming for is not just women's liberation but also a total liberation, a social, political and economic liberation. Our situation is very different to that of the European women. Look around you and see, especially in town. Hundreds of our women slave everyday for white women. They cook, clean, sweep, and wash shit for crumbs from European women. European women thought up Women's Liberation because they didn't have enough to do, and they were bored out of their minds. They wanted to be liberated so they could go out and work like men. They were sick of being ornaments in the house. They hate their men for it. That's not our position at all. Our women always have too

> much to do. Our women never have the leisure to be ornaments. Our societies are people oriented so we care for one another. Our situation also affects men. [cited in Jolly 1978, p. 52]

Similar criticisms of white, heterosexual, middle-class mainstreaming of feminist actions and agendas also held for working-class women or for women with special identities or citizen status (as migrants), in health status (e.g., disability), or sexual orientation. Lesbians, too, spoke quite openly, and for the first time in terms of sexual "citizenship" (Evans 1993, Plummer 1992, Stein 1993). We shall come back to this later in Chapter 9. Suffice it to say here that lesbianism was at the forefront of political dispute and for the first time introduced the notion of compulsory heterosexuality and of heterosexism. Sheila Jeffreys, in her book *The Lesbian Heresy* (1933), insisted on the political act of their identity:

> Before readers affected by postmodernism start to assume that such use of the word "lesbian" bespeaks essentialism it should be said that when lesbian feminists speak of "lesbian" anything, they are generally speaking of something that has to be consciously created by lesbians as a political act, not any natural "essence." Attacks by postmodernist lesbians and gays on lesbian feminist theorizing using male authorities such as Foucault and Derrida to back them up, should perhaps be understood as either willful misunderstanding or deliberate attempts to constrain the construction of an alternative lesbian vision. [p. 169]

Due to these unrests, feminism henceforth was committed to a philosophy of pluralism (Gunew and Yeatmen 1993). Despite the new consciousness (we are all different but we have common purposes and some common experiences), feminist theory has so far found no lasting escape or reprieve from deciding on the distinction between sex and gender. The sex/gender split is not just a feminist invention but to an extent is an inherited problem from contract theory (Bacchi 1991). The sex/gender split also entered the medical and psychological fields and, quite separately, it has been the theoretical point of departure for just about any inquiry of feminist theory.

Feminist theorists who adopted an identity politics of "difference" discovered a "new femininity" and "new masculinity" (Guillaumin 1985) and expressed deep concern about antiessentialists' denial of such intrinsic feminine identity (see Kaplan 1993). They held that the movement was centrally built around women and made no sense without an essentialist base. The aspiration of the early feminist movement was to create a political community of women in order to make visible their special civic demands. This political momentum would collapse unless it was on the basis of *being* women (Soper 1990). One may agree with Chantal Mouffe (1992) that denial of the intrinsic feminine identity does not lead to "collapse" for at least the two reasons she gives: (1) a critique of an essential identity does not lead to the rejection of any concept of identity, and (2) "the absence of a female essential identity and of a pregiven unity does not preclude the construction of multiple forms of unity and common action" (Mouffe 1992, p. 381). Arendt's view on identity politics is instructive here: Honig (1992) interprets her book *The Human Condition* (Arendt 1958) as saying that "a political community that constitutes itself on the basis of a prior, shared, and stable identity threatens to close spaces of politics, to homogenize or repress the plurality and multiplicity that political action postulates" (p. 227).

For those who argue that "gender" is a construction (as the term is actually meant to denote, in contradistinction to "sex"), it follows that laws ought to eliminate existing gender bias, at least as far as possible. By the designation of "gender" as an analytical category, this position is also criticized because it leads to a politics of sameness: women and men have the same potential and women are therefore entitled to the same rights and freedoms as men. In terms of policy making, the politics of sameness regards it as "dangerous" to even attend to women's sexual difference (Dietz 1991). In this context, formal political and legal structures become unconstitutional or untenable and, indeed, as Mezey (1990) points out, legal gender-based distinctions should be viewed in the same negative light as race-based distinctions. To quote Cornell (1992): "The wrong in discrimination is the imposition of rigid gender identities on sexual beings who can never be adequately captured by any rigid definition of gender identity" (p. 290).

Others argue, with justification, that "gender" needs to be overcome or be aided by other markers or concepts. Cornell (1992), for instance, spoke of equivalent rights and Burstyn (1983) extended the term *gender* to *gender-class* to imply oppression that she believes is conveyed neither by the term *sex* nor *gender*. Of course, the debate on other markers of social disadvantage (or even oppression) has now become widely accepted in feminist theory, but other writers quite rightly warn that the disadvantages produced by class, gender, and ethnicity do not always parallel each other and that a critical dialogue between groups is needed to increase feminist sensitivity to these disjunctions (Gunew 1990). Rhodes (1989) advocates reorienting law from a focus on "sex based difference toward a concern with sex based disadvantage" on the grounds that only in this way would analysis take into account "whether legal recognition of gender distinctions is more likely to reduce or reinforce gender disparity in political power, social status and economic security" (p. 1).

The debate has remained inconclusive. Zillah Eisenstein (1988) condemned the view that "difference equals inequality; sameness equals equality" as "phallocratic" ways of thinking and proposed a "completely pluralized notion of difference(s)" as the most appropriate foundation of radical egalitarianism (p. 100). Rhodes (1989) settled for the view that "the critical issue should not be difference but the difference it makes" (p. 83), and Mouffe (1992) ultimately defined a political identity rather than a sexual or gender identity when she turned the debate from identity politics to citizenship. Lynda Birke (1982) crucially saw that the perceived dichotomy between masculinity and femininity may have limited science. She also warned that the masculine/feminine dichotomy had been aligned with other dichotomies, often to justify imperialist expansionism and racism.

This is where one of the roads of feminist political theory has led and where we have arrived. Braidotti (1993) put it bluntly that "the paradox of feminist theory at the end of this [twentieth] century is that it is based on the very notions of 'gender' and 'sexual difference' which it is historically bound to criticize" (p. 29).

QUIET AT THE WESTERN FRONT?

The 1970s were a time of a politics of optimism. By the 1990s, this euphoria had largely gone; the women's movement had flagged; tensions, disappointments, and new backlashes seemed evident (Faludi 1992, Kaplan 1996). The time had come for more detached reflection and evaluation of the status of women. Tensions between this fear and remnants of an insistent optimism were more widespread in the 1990s than in the 1980s, when anger was not only often expressed but at last considered to be a legitimate emotion (Petrie 1987). At the turn of the century the mood changed to a more demure and tense posture, and some women braced themselves for changes as if foreboding ill winds.

For more than a decade, it was a widely held belief that women could "make the grade" if only they knew how. The idea was to get women into careers so that their working options would eventually carry them out of lowly paid, non-career jobs. Often built on a deficit model (women need to learn to be on par with men), a decade of teaching women in management and publishing social skills books followed. The politics of optimism spawned a glut of "how to win" books (Clutterback and Devine 1987, Davidson 1985, Scollard 1983) and informed women of the handful of outstanding women with successful careers (Wells 1987) so as to provide role models for future success. The discourse of optimism carried messages of success as just being around the corner, a thing worth battling for. While often very one-sided, middle-class, white, and elitist in tenor, it also contained egalitarian messages, resulted in affirmative action strategies, and promised to make the workplace a more equal, if not gender-neutral, environment (Burton 1991, Poole and Langen-Fox 1997).

By 1995, some claimed that policies had unduly favored women and any affirmative action had to stop. Apparently, women were "storming" the professions (Birrell 1995), to the detriment of men. However, the reality is somewhat different and on several levels: at the level of structural and indirect discrimination, on a personal level, and in the way women's lives may continue to be disadvantaged by the things they do and what may be expected of them (Poole and Langen-Fox 1997). There is evidence that women continue to do most of the housework even when they work in full-time positions or careers (Bittman 1991), except for some changes among the younger generations. A long list of publications from the late 1970s shows that women do not handle time badly but that they have multiple commitments and often conflicting time schedules and demands placed upon them (Apter 1985, Barron and Norris 1976, Beechey and Perkins 1987, Berch 1982, Gray 1983). Indeed, these socially created conflicts have not been eliminated but, rather, they have continued to disadvantage women by causing them to carry guilt about their inability to resolve them (Shaw and Burns 1993). There is nothing in particular that women, as a gender group, have done incorrectly in getting ahead or utilizing the opportunities seemingly before them. What has changed, however, are the goal posts for winning success and the rules by which to play.

In the last decade or so, these changes have undone more of women's progress than any direct discrimination against women; they have cemented a new and widening inequality (O'Leary and Sharp 1991) and a retraction of reproductive freedoms. The prognosis that women are slowly making headway into leadership positions in the private sector received a sharp correction by reports and surveys of the late 1990s showing that women in the United States and Australia occupied between only four to eight percent of directors' positions in companies (Carruthers 1997) or on Wall Street (Truell 1996, Valian 1999).

A new so-called third wave of feminism started in the 1990s, with a concomitant counterattack on feminism, this time by young women. In the third wave, some of the issues of the 1970s have remained on the agenda while many new issues have arisen. Eating disorders, self-mutilation, and body image form one important cluster (as shown by Wolf's *The Beauty Myth* in 1991 or Edut's *Adiós Barbie* in 1998). Drugs have been a major concern (Wurtzel 1997) and sexuality has had wide airing, ranging from discussion of child sexual abuse to HIV/AIDS, and including teenage sexuality (Tanenbaum 1999). Another cluster of interests has emerged around topics of technology, globalization, and equal access to the Internet (Baumgardner and Richards 2000).

The third wave feminism, despite repeated pronouncements by the media that the movements were "dead," has undoubtedly become a new subculture with a large clientele, particularly of young females. Indeed, guidebooks for girls have made a strong reappearance, reminiscent of the 1950s, although mostly with nonconformist content. The subculture, however, has left the intellectual establishments largely untouched. Indeed, Katha Pollitt complained in 1999 about the number of self-absorbed books on women's condition (see Baumgardner and Richards 2000, pp. 18–19). However, this is not the entire focus of the new feminists; many third wave writers have directed their energies to social and political issues of immediate concern.

At the same time, there are probably more women today than in the 1970s, from Katie Roiphe (1993) to Danielle Crittenden (2000), who embrace an essentialist position and believe in innate differences between the sexes, and act as colluders, whether knowingly or not, in the influential new politics of biology. There is relatively little issue taken with evolutionary psychology despite or because of the fact that the latter is thriving in elite institutions, especially in the United States, and increasingly also in Britain with essentialist and deterministic claims about gender and ethnicity (see Rose 2000). It is surprising that so few have argued against it, given the fact that the essentialist debate is extremely far-reaching because it has, not just as a byproduct, once again reified old values and beliefs about the different "nature" and alleged "innate" roles of women and men.

However, those feminist scientists and biologists or theorists of science who have taken an active role in entering the debate on the use and abuse of genetic argument have been extremely productive and innovative in recent years and some of their writing has semi-cult following (such as Haraway's [1991] "Cyborgs"). The writings of these women are cited throughout the text because, quite often, they have been at the forefront of a critique against reductionist and deterministic thinking. (For an excellent bibliography see Rose 1994.) The year 1978 was a watershed for women in science when *Signs* published a special issue called *Women, Science and Society*. In this domain and constellation, women have innovatively problematized the role of science in society, and their own role in science. However, it is also important to point out here that such intellectual activities have earned them enormous scorn from the new believers in genetics, who have dismissed their concerns as mere political correctness or as ideological whimpers in the world of hard science (as we will mention several times again).

Evolutionary psychologists tend to be extremely angered by any criticism against them on the basis that they have been misunderstood, misquoted, and their writings distorted. So they tend to enter a plea of innocence to any charges brought against them (Kurzban 2002). In a manner of speaking that plea of innocence may well be justified, but not in their social translation. Much of their work and many of their authors have shown more than just a tinge of wishing to seek influence in the social arena on the basis of hypotheses that have, at times, no more claim to fame than speculation. The resurrection of the genetic basis of difference in men and women is an old idea in a new garb.

6

Women's Biology and the Consequences

WOMEN'S BIOLOGY AS DEFICIT

Only relatively recently have women been granted control of their own sexuality (without being considered evil) and allowed access to universities (without being considered abnormal). Both changes happened at about the same time, toward the end of the nineteenth century. Although they seem poles apart, in fact, they had a great deal to do with each other. So long as women were considered to be driven by their sexual desires and passions they could not be allowed entry into universities (where competition was based on rational thinking) or take professional roles in public life. Biology provided most of the arguments against women's wishes to be educated and enter the professions.

Hence, when women began to demand entrance rights to universities, a heated counterdebate arose around the Darwinian theory of the evolution of sex differences. One of the issues was whether female brains were smaller, and thus had less intellectual capacity than male brains (Fee 1979). Assumed "proof" of women's biological inferiority to men eventually had to be discarded when scientific measurements of cranium size showed it to be incorrect (see Salzman 1977). Another argument, promulgated by Herbert Spencer (1864–1867), and a much more dangerous myth than that developed by craniologists, was the assumption that women could "overtax" their brains and thereby "cause" themselves to have diminished reproductive ability (Peel 1972). A reiteration of this view, very persuasively argued at the time, can be found in Clarke's disastrously successful book *Sex in Education* (1873), which haunted at least two generations of girls and was still around after the Second World War (discussed in Sayers 1982, p. 9). Yet another argument, now known to have no basis in biology, concerned sexual dimorphism in metabolism. Women were supposed to have a different and inferior form of cell metabolism that caused them to be sluggish, passive, and less able to study. Exactly the same argument was used later by white people to "explain" the perceived inferiority of black people.

When women began to demand their right to enter the business world at levels of responsibility and leadership in the 1960s and 1970s, another set of biologically based arguments appeared that were readily usable against career aspirations of women. Women's behavior was said to be controlled by the ebb and flow of their hormones, and that during a phase, said to occur premenstrually, their behavior became erratic and unreliable and their intellectual capacity decreased (e.g., Dalton 1979; see Vines 1993 for a critique). Thus, it was argued, even if a woman had numerical or mathematical abilities *equivalent* to those of men she could not be employed in a position of responsibility using those abilities because her performance would deteriorate for a period of at least four days every month: "You would not want the president of your bank making a loan under the raging hormonal influence of that particular period" (E. Berman 1970, p. 35). A similar argument was used, also in the 1970s, by commercial airline companies in Australia to prevent women from becoming pilots.

Symptoms listed as being associated with premenstrual syndrome (PMS), also known as premenstrual tension (PMT), range from mood disturbance to crashing planes and committing crimes and suicide but, although PMS is claimed to be a medical entity with a hormonal cause, there is no consistency between studies as to the time at which it is said to occur during the menstrual cycle (Parlee 1973, Rogers and Rogers 2001). There is also no consistent collection of symptoms that occurs in any one woman. Noticeably, most studies that have looked at behavioral changes over the menstrual cycle have failed to consider that social attitudes may be important, and ignore two studies showing that education about menstruation prior to the age of menarche has a significant effect on changes in behavior at different stages of the cycle (see Rogers 1979). Yet, a rigid belief in hormonal determination of an ill-defined condition, given reality by name only, still persists. Moreover, although it may be recognized that far from all women suffer from PMS, the alleged existence of it has been applied to all women, presumably to raise doubts about women's suitability for certain professions. Sociobiologists writing in the 1970s and early 1980s, including Wilson (1975), Trivers (1978), and Tiger and Fox (1978), as well as psychobiologists Bardwick (1971), Buffery and Gray (1972), Hutt (1972), and Levy and Gurr (1980) all argued that the differences in behavior between the sexes are determined by genes and hormones. The question they all posed was: What was the use of implementing the changes that feminists were demanding in the face of the evidence that the inferior position of women was anchored to their biology? "Scientific" publications purported to give reasons for resisting social changes.

SEXUALITY VERSUS PERSONAL FREEDOM

Women's movements of various shades of political conviction also tried to tackle social and psychological issues regarding female sexuality. They generally understood that a woman's freedom of movement was strongly dependent on the perceptions of her sexuality. The first bourgeois women's movements (1880s) argued that women should be granted full rights because both women and men were essentially the same human beings. Yet they also stated firmly that women were essentially different from men, and in some ways superior to them, and that, since the two sexes were complementary, women had to be granted rights so that society could function successfully (Adams 1988, Kaplan and Adams 1990).

This claim derived from a belief in innate biological differences between the sexes, the main distinguishing characteristic being the female instinct to nurture. The belief in innate differences between men and women developed along with the Victorian cult of domesticity, which removed women from their traditional stereotypic role as lascivious tempters of men, and instead placed them on a pedestal of moral purity, with the task of providing men (now the ones driven by uncontrollable instincts and lusts) with spiritual uplift. By the late nineteenth century, this view became linked to Darwinist assumptions that certain instincts, values, and character traits were both sex linked and biologically determined. Such claimed scientific "evidence" gave support to the notion that separate male and female roles were natural and immutable (Bridenthal et al. 1984).

In the movements of the 1900s, it was believed that women's superior morality had to become the standard for both sexes, creating something "finer and deeper." The arguments were directed toward ending men's domination in sex and marriage because it caused "a painful annihilation of [women's] strongest and most inner life instincts" (Bäumer 1904–1905, p. 326). However, having argued in this way, women were then forced to deny any sexuality of their own, even their sensuality, and to proclaim only its sublimation into "motherly" love. Campaigns were held against prostitution and sexual harassment and for reform of family law. These rested on a conviction that the double sexual standard oppressed women and that women's powerlessness in the private sphere of the home was connected with their exclusion from the public sphere (Adams 1988).

The social consequences of the nurturing image were enormous, at least in terms of the personal freedoms it immediately gained women (including admission to universities). If their sexuality was not a rapacious drive to conquer (as allegedly was men's) but rather more diffuse and linked to traits of altruism and nurturance, most typically seen in motherhood, then single women could also act according to their natures. They could remain unmarried without being unnatural women by carrying out tasks of "social motherhood." This explicit permission given to single women to enter the public sphere was eagerly taken up by many before the First World War (Slaughter and Kern 1981). It not only justified their careers, but also allowed them to postpone marriage indefinitely and to enter into close, intimate, and even long-term relations with other unmarried women, usually without raising eyebrows from contemporaries.

However, personal freedom was obtained at the price of having to hide and even deny female sexuality and, to some, this was an entirely unsatisfactory state of affairs. In the 1920s some feminists interested in sexual issues joined with a group of male sexual researchers and announced, as if it were entirely novel, that women actually did have sexual desires (Bridenthal et al. 1984). They believed that women should have the right to seek sexual fulfillment and that this should include orgasm. That radical position (at the time) was substantially modified, however, by being centered on heterosexual intercourse with penetration. Although women were granted recognition as sexual beings with a right to sexual pleasure, it was on male terms and on heterosexual grounds. Female sexual responses were seen as supportive of male dominance (Grossman 1983). Men were encouraged to take the active role in sexual matters, initiating and teaching their women, who were either passive or resistant (in which case they had to be subdued). In addition, this new orgasm consciousness led to new stress for many women in sexual relations. Women who did not respond were labeled abnormal, frigid, or lesbian. Those unable to reach orgasm now had either to "confess" when they did not, and risk being labeled frigid or unfeminine, or to fake it. In the case of the former, new unhappiness resulted for both partners (he was not the man he thought he was; she was unsuccessful in her most womanly role). In the latter case, a new dishonesty crept into the most intimate relationships (cf. Grossman 1983, Kaplan and Adams 1990).

Enjoyment of heterosexual intercourse rather than the earlier diffuse notions of caring, mothering, and bonding, was now emphasized. Despite the costs mentioned above, many young feminists found the message of the sexologists personally appealing and, since this message was coupled with a new emphasis on contraception and abortion, also liberating. The flapper could not only dress with less restriction and go out on her own, but could also enjoy sexual relations. For the young professional women particularly, this new reform agenda seemed to grant new independence to women in their relations with men (Bridenthal et al. 1984).

However, the views of the sexologists challenged the very basis of some key feminist claims of the pre-war era. Their analysis denied single women the right to remain single and to have close emotional and/or erotic ties to other women. "Normal" sexuality was now linked to male-centered intercourse and was confined to marriage or heterosexual partnerships. This, then, undercut the feminist argument of social motherhood, for the only model offered by the reformers was that of the heterosexual woman planning real motherhood. Then feminism shifted to emphasize a depoliticized, individualized role for women in which the single woman no longer had a valid social role in civic or political life, while "normal" women gave primacy to sexual relations and motherhood.

Under fascism, especially under National Socialism in Germany, debates about women's sexuality changed yet again. In Nazi writings, the mother was considered to be the source of the race. A note of urgency entered these writings, a plea for speedy actions and programs to prevent further contamination (biologically, sexually, morally), further decline, and even the death of the German people.

Nazi authors were careful to distinguish between deserving and undeserving mothers, a notion that had already become commonplace in the eugenics and sex reform movements (Grossman 1983). As we mentioned before, this notion has reappeared in debates of population control in the present time and is openly advocated as "biopolitics" (Cattell 1987, Miller 1997). The Nazis later realized their plans for eugenics with their forced sterilization program (Bock 1984, 1986). Sexuality, of females in particular, was reinterpreted as evil, foreign, irrational, and therefore dangerous. Jews were said to have a strong sexuality; they were seen as capable of luring innocent German "maidens" into a moral and genetic abyss. To Nazis, Jews represented a "dangerous pollution" of the gene pool and sexuality (as desire); they could only lead to poor decisions, the most disastrous of which would be actions against the "bright" destiny of "rejuvenating" the Germanic peoples (Kaplan 1994).

In Nazi ideology, the first marital duty was *sexual* fidelity to a single partner and the second to bear eugenically perfect children, because only in this combination were there bound to be positive consequences for the race. The fidelity was not perceived to be sexual, but rather became disembodied into a compatriotic quality. The policies of the women's movement, Nazis argued, had resulted in a devaluation of morality and motherhood (Bock 1984).

The new fascist opinion makers (women among them) campaigned against abortion laws, the advocacy of contraceptives, and sexual pleasure, asserting that they were apparently extremely harmful to women.

In their view, the separation of sexuality and reproduction was "tearing apart women's souls" and creating a perilous situation in which women failed to continue to guard the innermost values of the *Volk* soul. Indeed, personal feelings of shame and modesty were regarded as a biological necessity for reproduction, because lust resulted in women losing their motherly instinct (Kaplan and Adams 1990). Alfred Rosenberg wrote in 1930 in *The Myth of the Twentieth Century*: "Emancipation of the woman from the women's emancipation movement is the first demand of a female generation that seeks to save *Volk* and race, the eternal unconscious, the basis of all culture, from perishing" (cited in Mosse 1978, p. 66). Male Nazis also wrote on female sexuality and the link between sexuality and motherhood. In his book *Sexual Hygiene* (1939), published in Berlin and widely read, Max von Gruber wrote that

> if sexual intercourse is practised from the start only for the purpose of pleasure, it poisons the relationship of the spouses to one another and in particular harms the morality of the woman. She will no longer view the execution of intercourse, as nature inclines her, as an act of awe and meaning and of great consequences in which the mysterious primeval forces of life are the hidden driving power, but she will gradually learn that it is nothing but a pleasure. [p. 101, transl. GK]

Klaus Theweleit (1987) has since shown in relation to proto-Nazi men of the 1920s that their positive image of women was limited to those who were mothers or sisters, "desexualized creatures who related to men within the constraints of kin, with no threatening erotic femininity" (Ch. 1). Nazi women publicly proclaimed their wish to sacrifice themselves, implying that sexual enjoyment as such was reprehensible. Their fear of turning into sex objects or becoming victims of their own feelings made them accept an image of womanhood as reproductive machines, for the sake of a eugenically improved race of sons.

The sexual revolution of the 1960s in the Netherlands and the 1970s about everywhere else in the Western world (Kaplan 1992) reinstated women's sexuality and theoretically gave women the freedom to pursue careers with no further expectations of motherhood as the only (or only true) fulfillment of womanhood. The sexual revolution also resulted in an extremely rich field of writings on women, their bodies and desires (Coward 1984, 1985, Grosz 1989, Haug 1987, Irigaray 1977, Rich 1976).

Arguments have been detailed here to indicate how many social and political decisions hinged on the interpretation of the nature and extent of women's sexuality. The views voiced in the first half of the twentieth century have not, unsurprisingly, disappeared. After all, sexuality can be described in many ways, from the erotic to the orgasmic, from the diffuse to the passionate, from the sexual abandon for pleasure to the intentional reproductive act of intercourse. With hindsight of more than a hundred years of intense sexology studies, the information is hardly new.

What is remarkable is the relationship between women's sexuality and the granting of specific freedoms in civic life. Such questions were never asked of men, nor can we recall a single theory that proposes that only a diffuse and fatherly sexuality (nonorgasmic) would be a suitable precondition for men to enter public life. Moreover, when woman's sexuality and her "true nature" had been defined and her relationship to her own body and sexual desires made clear, it was almost always with some sense of deficit and punitive social consequence, whatever the theory. The supposed existence or nonexistence of female sexual desires resulted in social costs in each case. Evidently, biologically based arguments are so malleable that they can provide proof for biological opposites with equal ease. The only consistency in this charade of anti-female arguments is the strong evidence that female independence and freedom had to be prevented at all costs.

Sexuality has been debated with renewed intensity since the Kinsey Report of 1948, and it has held center stage in the theoretical and empirical work on social organization of gender relations ever since (for a recent review see Lorber 1999). Gender organization of sexuality is not a new topic but it is a relatively recent phenomenon (cf. Rubin 1984) that sexuality and gender have finally been analytically separated and discussed as separate, albeit reinforcing, concepts (Nakano 1999), as we will discuss in Part III. Breaking the nexus makes it possible to work analytically with the term *gender* as a social construct (as it is actually meant to be used) without tying women to their biology and without enabling those with vested interests to derive socially oppressive rules from the possession or absence of active sexuality.

WOMEN AND THE IQ DEBATE

The first round of admissions of women to centers of higher learning was intimately tied to the perception of their sexuality. Stepping into the public arena was partly possible because women were not seen as sexually dangerous, as we said before. That concerned only the admission to civic life. It said nothing about men's perceptions of the intellectual abilities of women. Achieving admission to universities and gaining the right to participate in education and in some employment was just one step. Each step that a woman took on this road was on a slippery slope of male suspicion and prejudice because it was "unnatural," exceptional, and, ultimately, perhaps, doomed to failure. This perception, emanating from understandings of biological sex differences, led researchers to take up unexpected tasks. It resulted in disproportionately large research efforts throughout the twentieth century to establish beyond a shadow of a doubt that women were tied to their biology and that sex differences in brain organization and function placed men ahead of women. In other words, a good deal of that scientific research did not aim at impartial outcomes.

Although by the turn of the nineteenth century F. P. Mall (1909) had demonstrated that there was no evidence for a sex difference in the gross anatomy of the male and female brain, this by no means quelled notions of the inferiority of the female. Havelock Ellis formulated the theory of "greater male variability." He believed that, whatever characteristic one measured, there would be more variability in a group of males than in a group of females. This variability referred to a wide number of physical characteristics, ranging from size of the knee or hip joint to height and weight. We do not know how valid his original data actually were, and to our knowledge they have never been put under close statistical scrutiny (see Anastasi 1966).

Yet Ellis's obscure theory acquired unjustified limelight when it was applied in the 1970s to interpretations of IQ tests in men and women. In her book *Males and Females*, which was part of a rekindled interest in sex differences (and racial differences), Hutt (1972) attempted to explain the inferior position of women in society by adopting the theory of faster development in females. She also resurrected the idea of greater male variability, an observation she now referred to as a "law," and extended Ellis's claims to propose that IQ scores obey the same principle of greater anatomical variability in males. To substantiate her claims, she referred to only one study of IQ, the work of Heim (1970), who proclaimed the mediocrity of women on the basis of his finding that on IQ tests males have the lowest and highest IQ scores, whereas females are more closely clustered around the mean. This explained, for Hutt, why more males are great artists, scientists, and businessmen, the reason being that more men than women have exceptionally high IQ scores. The history of exclusion of women from the universities and professions was ignored and the cause of male success in the arts and sciences placed firmly on a biological basis. Interestingly, Hutt also said that more men than women were said to be in the very low range of IQ scores but this did not receive any further attention. The scores of women, according to Hutt, were closer to the average than those of men. This is another way of expressing female inferiority, now in terms of mediocrity, or being more average than men. According to Hutt, and to those who think likewise today, the greater achievements of men relative to women in public life are a consequence of their genes and hormones and not men's greater opportunities now and in the past. To quote Hutt (1972, p. 90): "The fact that males predominate in the intellectual and creative echelons seems to have a basis other than masculine privilege." There were other pseudoscientific reasons given for male intellectual supremacy as well.

Also in 1972, another research team stepped forward with another explanation for male superiority. Money and Ehrhardt proposed that testosterone affects the brain of the fetus during development so that the individual exposed to testosterone is later more dominant and has a higher IQ, along with a number of other behavioral attributes. Their widely publicized data apparently showed that females exposed to abnormally high levels of testosterone during fetal development have higher than average IQ scores, are more tomboyish and more dominant. These findings, based on a small sample size with inadequate controls, have not been substantiated (see Rogers 1981). Geschwind and Galaburda resurrected this dubious hypothesis in 1987 with no further evidence to substantiate it either. They hypothesized that abnormally high levels of testosterone during fetal development causes giftedness in males. Females rate little direct mention in their writings.

Even at the turn of the twentieth century, a hundred years since women's first admission to universities, there were still some voices arguing that there exists a biologically based inequality reflecting a "natural" order and hence any difference in social status between men and women is not really unjust. Doreen Kimura (1999) offered the explanation that unequal representation of women and men in professions requiring spatial skills is due to the "fact" that women supposedly have inferior spatial abilities than men. This, she believes, is a reflection of the biological differences between women and men rather than an aspect of social inequality and the result of social perceptions that perpetuate unequal representation of women and men in the professions. Others, who likewise wish to argue for *general* sex differences in intelligence between men and women, adhere to the idea that IQ measures a general intelligence, labeled *g*, which relates to particular general aspects of cognition, such as processing speed and working memory, features that are said to be inherited (Luciano et al. 2001). Still others are prepared to say that *g* had an adaptive significance for our ancestors and even that it might have been subject to sexual selection, meaning that it was a way by which males attracted the opposite sex and competed among themselves to do so. According to this idea, men therefore faced selective pressures that enhanced their intellect (see Mackintosh 2000).

The problem is that writers in science who are focused and determined to find biologically based differences in intelligence may not always have read widely and may not be aware of the history of women's outstanding intellectual achievements, made when they were given the opportunity. History books have at least recorded them, even if scientists have not read them. For instance, to quote just one passage from an account of the Italian Renaissance:

> During the Renaissance women distinguished themselves not only in prose and poetry, thereby proving their equality and sometimes even their moral superiority to men, but they filled with distinction several of the most important chairs in the universities of Italy. . . . At the epoch when the famous medical school of Salerno flourished, in the fourteenth century, Abella, a female physician, acquired a great reputation in medicine and wrote books in Latin, and in the following centuries women professed the sciences and the classics in the universities of Bologna, Brescia, Padua, and Pavia. It is during this period that we encounter the names of Bettina Gazzadini, of Mazzoloni, whose bust is found among those of the most distinguished savants in the anatomical museum of Bologna; Laura Bassi, who lectured on physics in Latin; Marie delle Donne, Elena Cornaro,

> Clotilde Tambroni, and the celebrated Gaetana Agnesi, whom no women and few men have ever equalled in the extent and profoundness of her learning and in the goodness of her heart. I shall not speak of the beautiful Novella di Andrea, of Laura Cereta-Lerina, at twenty professor of metaphysics and mathematics at the University of Brescia; of the celebrated Neapolitan, Martha Marchina, professor at a German university, an honour enjoyed at that epoch by Italian women alone; of Pellegrina Amoretti, of Isotta da Rimini, and of so many other heroic souls who, toiling in isolation and surmounting unheard-of difficulties, ever protested with word and pen against the inferiority attributed to their sex, and finally won the admiration of men. [Stanton 1884, pp. 312–313]

If we made a list today of women in the arts, science, politics, and business, we would fill volumes and it would be exceedingly difficult to argue that the quality of these achievements lags behind those of men. The point in science is that any theory needs to be verifiable and stand up to scrutiny. If there are examples that defy the assumptions then the idea cannot be upheld. Since the empirical evidence for the above arguments has been getting progressively weaker over the past few decades the scientific claims for a genetic basis of sex differences must further be regarded as severely prejudiced. Those who argue that women have been given opportunities that they did not deserve, as some believe, and have been pampered along and promoted well beyond their capabilities (a point that is also made with respect to men and women of color in Western societies), we can only reply that the history books, autobiographies, and detailed statistical analyses are full of examples of women having to take many more hurdles than men to comparable positions (Jennett and Stewart 1987).

It is common to measure and plot IQ as if the intelligence of an individual could be encapsulated into a unitary score determined by a single test. There are many different forms of intelligence and the scores obtained from just one IQ test do not tap all of them. The result of plotting the distribution of IQ scores for the population is a bell-shaped curve. Yet this is not merely a statistical exercise because society makes decisions about the education and future of individuals whose IQ scores fall on the extremes of the bell-shaped curve (that is, considered to be abnormal in a statistical sense). Those with scores falling in the low range are excluded from normal schools. Those with scores in the high range are said to be gifted and are often allowed certain privileges in education.

IQ measures are based on dubious assumptions in more than one way. It is assumed that IQ tests are a measure of intelligence when they might well be an indication of cultural proficiency. The more integrated into the dominant culture an individual is and the more opportunities he or she has had, the better the performance should be. Within limits, this generally holds. People from the middle class score higher than from the working class, and those from mainstream culture perform generally better than those in the cultural margin. So, blacks tended to perform at lower levels than whites, migrants equal to blacks, and women, in the past at least, at lower levels than men. Some behavioral geneticists and psychologists thought that they could derive insight from such results by hypothesizing essentialist categories based on certain groups (at the population level) being more or less "intelligent" (Eysenck 1971, Jensen 1969).This assumption, in turn, is based on the idea that intelligence is hard-wired into the genes, implying that one's intelligence is fixed. However, IQ scores can change with age, either increasing or decreasing depending on experience (Bowles and Gintis 1976).

The reason the commonly used concept of "potential" is quite misleading is that it assumes that intelligence is like a pot with goods in it and the pot may be larger or smaller for one or the other individual, but one can ultimately only use what is in the pot. This is a very static view of intelligence. Performance on IQ tests also depends on motivation, rewards and punishments, social pressures, and a multitude of other factors that may be rather intangible.

There is thus an assumption that the gene pool in a given population will ensure the same production of "intelligence" and that production cannot be altered. Eyssenck and Jensen's idea in the 1970s could largely be summed up in one sentence: we are white, so we are bright. Further, we have always been bright and will remain so. Evidence actually throws major doubts over these simple equations. First, there is plenty of evidence that people in marginal groups perform at higher levels once they move to the center. Second, IQ scores, according to the standards set by Western societies, fluctuate. In 1987 and 1994, Flynn published a series of results from his historical surveys involving measures over a sixty-year span and in fourteen nations. He found that the population mean of IQ had risen substantially in that period, by as many as twenty points or more between the 1920s and the 1980s. For the United States, he also found that the black population now showed

IQ values that were the same as those of the white population in the 1920s (Flynn 1994).

These results are not explicable if one assumes genetically fixed intelligence, but they are understandable in terms of social, political, cultural, and also biological influences. The period of time under investigation happens to coincide with the greatest expansion in the education systems in the history of humanity, particularly in the tertiary sector. It also coincides with a new set of values that, while present before, acquired an unprecedented status at the beginning of the twentieth century—a cultural value of innovation (expressed in grants and awards for innovative ideas, patents for new products, and so forth). Another social value is "change." Change requires accommodation and a context in which this can be achieved. Moreover, nutrition may play a role in raising IQ values. In the Western world at least, famines have disappeared from the calendar of regular "natural" events. The inhabitants of the first world also eat better food and do so more regularly for a larger percentage of the population, and from an early age on.

The question of variability of IQ scores has another possible flaw that is well known but not always acted upon. Sex differences in distribution of IQ scores, to a considerable extent, can be a result of how the tests are constructed rather than a reflection of any basic difference between women and men. Let us give the example that women and men may perform differently on various questions in an IQ test, men perhaps performing better on spatial ones and women performing better on verbal ones, by no means because these differences are determined by their genes. To give a simple example: if two-thirds were questions on spatial ability and one third on verbal skills, the results could be skewed strongly in favor of males. This may be too obvious a bias but a research bias of this kind is administered regularly to different cultural groups. The dominant cultural group assesses the abilities of a wide range of cultures (in multicultural societies) without adjusting questions in favor of cultural differences. Similar (and often subconscious) scorer bias may be at work in tests designed to establish sex differences between male and female intelligence. The overall IQ result can thus depend on the balance of questions in the particular test, rather than being a manifestation of something in the biological nature of the person being tested. There have been IQ tests on which women score higher than men, but these have been adjusted to eliminate the apparent female superiority. Adjusting the questions asked in an IQ test may equate the mean scores of men and women but leave the distributions differing, as Hutt (1972) noted to be the case for just one such test. Another balance of questions might match the distributions of scores for males and females but leave the mean scores differing.

In fact, the differences in the distributions of overall IQ scores for women and men vary from one test to another. In addition, there is evidence that IQ tests are not as independent of past experience and social position as is often claimed (Bowles and Gintis 1976). This might also explain at least some of the differences in the scores of women and men. To dwell on the sex difference in the distribution of IQ scores, and in particular to draw attention to the high end of the scale to explain why more great scientists, artists, musicians, and so on are men, is not value-free but rather an aspect of negative attitudes toward women. IQ tests are designed to predict performance in different contexts and for different purposes and, as such, they are not measures of any absolutes of an imputed intelligence. If the dominant paradigm happens to be white, middle-class, and male, as it has mostly been, any colored, lower-class, female, or culturally foreign person may have less favorable results. These are then taken erroneously to be proof of lower intelligence and, in turn, the reason for their lack of economic success, as pointed out many years ago (e.g., Bowles and Gintis 1976, Rose 1976).

McGue and Thomas Bouchard (1998) have presented evidence indicating that genes may influence a person's choice of his or her social environment and so predispose the development of certain mental and other characteristics, and suggesting that this might influence IQ results. But this does not appear to be the case. If genetically determined sex differences in thinking and behavior existed, some degree of difference would remain even if both sexes were given exactly the same opportunities to develop their abilities and interests. There is evidence suggesting that this is not likely to be so. Alan Feingold (1988) examined sex differences in the performance of school children in the United States on scholastic aptitude tests from 1947 to 1983, an important period of change with respect to differences in social treatment of gender in education. The usual sex differences were found: girls did better than boys in verbal abilities and boys did better than girls in spatial and mathematical tasks, but these differences, to quote the researchers, "declined precipitously" over the years surveyed. That is, the recorded sex differences were greater in the early studies than in the later ones. Another study by Janet Hyde and colleagues (1990) found the same result as Feingold. They reviewed a hundred published reports of studies on sex differences (they used the term *gender differences*) in mathematics ability and compared the size of the differences in studies published before 1974 with those found in studies carried out after that date. The sex difference after 1974 had declined to half of what it had been before. It is not hard to guess that the decline in differences was not the result of a sudden evolutionary leap of females to become substantially brighter but a product of changing attitudes to female career aspirations and a decrease in gender stereotyping of careers.

So it is most unlikely that sex differences in spatial, mathematical, and verbal performance are inherited and in this way immutably built into the biology of girls and boys. Instead, they are manifestations of social values held at a particular time. Boys and girls are not driven by their genes to select particular learning environments and to develop particular abilities that alter their performance on IQ tests. In other words, the genes do not lead an individual to focus attention on learning sex-typical behaviors or developing sex-typical abilities. If girls and boys are given equal access to all forms of education and the expectations for them to perform in sex-typical ways are reduced, so are their gendered presence in sex-typed educational fields reduced. Feingold's (1988) results show that, when they are allowed to choose what they will learn and there are fewer pressures from society to conform to the gender-typical image, girls do not actually seek to learn skills that will enhance their linguistic performance, which might then make their abilities superior to that of boys. Similarly, boys also do not seek to learn those spatial and mathematical skills that would make their performance better than that of girls. He concludes that the social environment, not the genes, has a direct effect on the development and expression of at least some of the characteristics that have been considered typical of girls or boys.

SPATIAL ABILITY AND PERCEIVED SEX DIFFERENCES

Many researchers tend to treat studies on sex differences with some degree of finality, as if something was proven forever and an eternally valid difference had finally emerged. We have already noted in Feingold's results that a different social and educational climate over time made sex differences in spatial ability all but disappear in the younger generations. In addition, Paula Caplan and colleagues (1985) found that there are conflicting results on sex differences in spatial ability from one study to another. After reviewing the scientific literature, they concluded that there were no grounds for saying that men have better spatial abilities than women. Furthermore, there are cultural differences in the relative performance of males and females on spatial tests, and, for example, the Eskimo culture breaks any paradigm of male superiority. Eskimo women have been shown to have better spatial abilities than men (Berry 1966, 1971). This exception tells us something very important: sex differences in spatial ability are not a biological universal that evolved in ancestral humans; cultural factors of some kind can even reverse the direction of the difference.

Yet, despite these indications that there is no simple or universal sex difference in spatial performance, evolutionary psychologists, ignoring contrary results, have formulated a general theory claiming a genetic basis for male superiority in spatial cognition, mediated by the action of the sex hormone testosterone on the brain (e.g., Hampson and Kimura 1992). Together with colleagues, Kimura tested women and men on a task requiring them to read a map and memorize a particular route. The men learned the route in fewer trials than the women did but, once they had learned it, the women remembered more of the landmarks than did the men. The researchers interpreted this result as showing that testosterone causes superior navigational abilities, and then they linked it to the male evolutionary past, when there was a requirement for males to use these spatial skills in hunting. Whether the map reading ability tested by Kimura and colleagues has anything to do with the skills used in hunting is pure speculation, and to think that ancestral men performed these tasks better than ancestral women is moving speculation into fantasy. The fossil and archaeological records of ancestral humans tell us nothing about any differences in the spatial abilities of women and men.

It is possible that, along with other factors including learning, one or more of the sex hormones can modulate performance on spatial tasks, but their interactive effects are not simple. A study by Valerie Shute and colleagues (1983) shows some of the complications. These researchers found that men with low levels of testosterone performed better on spatial tasks than did men with high levels of testosterone. In the same study women with higher levels of testosterone performed better than women with lower levels of testosterone. One might conclude that there is an optimum level of testosterone, such that levels higher or lower than the optimum impair spatial performance, but it is uncertain whether this would provide an explanation for any sex difference in spatial ability, since many women have higher levels of testosterone than many men. This large overlap of testosterone levels between the sexes means that it would be very difficult to tease out any sex difference in spatial ability caused by testosterone. In other words, the sex difference itself is likely to be largely, if not solely, caused by factors other than the action of testosterone on the brain. Also, if the circulating levels of testosterone do affect spatial ability, the other sex hormones, estrogen and progesterone, might also have some effect. Some evidence suggests, however, that women may perform best on spatial tests at the stage of the menstrual cycle when the levels of both these hormones are lowest (McKeever 1995). Clearly, the balance among all of the sex hormones would need to be taken into account, but it is rare for any one study to consider more than one hormone at a time. Cortisol is another important hormone, also a steroid like the sex hormones, which acts on the hippocampal region of the brain known, from studies of animals, to control performance in spatial tasks. Given these facts about cortisol, it is reasonable to suggest that it may have a role in the hormonal mixture that affects spatial cognition. If so, the picture would become more complex but also more interesting, since cortisol is a stress hormone and the stress levels of the subjects being tested would have to be taken into consideration. For example, in a comparison of performance of any two groups (e.g., women versus men) we would need to assess whether the results are influenced by differing levels of stress in the two groups at the time they were tested.

We said earlier that along with changing attitudes to gender taking place in Western society, there has been a decrease in the measured differences between girls and boys on spatial and mathematical tasks, but the ability to match figures rotated at different angles has remained unchanged during the past decades (Masters and Sanders 1993). On such tasks, the greatest sex difference in performance occurs when two sets of figures, rotated at different angles, have to be matched in three-dimensional space. Interestingly, in this task, males are faster in identifying the match. This result might perhaps indicate a universal genetic and hormonal cause of this aspect of cognition, and this is worth further investigation, but it could also be that, so far, any changes in the way in which we raise girls and boys have been insufficient to affect this particular trait. The mere fact that a sex difference exists widely and persists over generations is, in itself, insufficient evidence to say either that genetic or hormonal factors play the major role in causing it, or that there might be grounds to trace its purpose back to something in our evolutionary past.

Claimed Relevance of the Human "Hunter–Gatherer Past" to Spatial Ability

Evolutionary psychologists, including Pinker (1997), Tooby and Cosmides (1992), and the popular writers Anne Moir and David Jessel (1993), argue that hunting has been a major force in the evolution of the human mind. This is alleged to be relevant to spatial ability and a host of other traits. To be able to form mental maps of where things are and how they might change with the time of day or the seasons would be required of a successful hunter. To find food, they argue, males had to travel further from the home base than females, and to hunt effectively they had to be able to throw at objects with accuracy. Since both of these behavior patterns depend on spatial cognition, they then argue that men evolved better spatial skills than women. Females, on the other hand, stayed in the camp or wherever they lived, caring for the children and engaging in verbal exchange. They were the homemakers. We note that this description of women as homemakers was typical of nineteenth- and twentieth-century Western societies with their ideals of domesticity for women. Having made the assumption that men were the only hunters and that women would not have needed spatial skills to locate and gather vegetables and fruit, the sociobiologists and evolutionary psychologists then tell us that modern humans have retained the same genes that lead to the assumed division of labor between men and women. By this series of assumptions, male behavior is given superiority and then linked to superior cognitive abilities. A similar argument had been made to claim a genetic basis for sex differences in aggression, as we have already mentioned.

Through a process called sexual selection these designated roles, it is said, led to sex-linked genes for aggression in males. These are still expressed in men in modern society. Here further assumptions are added on since these postulated genes for aggression are said to explain, and justify, male aggression from the football field to nuclear war. Again, Moir and Jessel (1993) state the sociobiologists' position in a nutshell:

> For most of our past, we lived in communities which depended for their very survival on hunting animals and gathering plant food. Men, with their greater strength and stamina, their roving tendency, and their greater skill in relating the spear to the space occupied by the prey, did the hunting, an unpredictable and dangerous activity. Women gathered nuts, grain and grubs—a safer and surer pursuit. [p. 132]

Alternatively, some argue, male wanderings might have brought more success in reproduction. The hypothesis of male range size was considered in the context of animals. It states that males increased the area over which they roamed so that they could encounter more females and thereby increase their mating opportunities (e.g., Barash 1979). In turn, the hypothesis states, this might have led to increased spatial ability and enlargement of the part of the brain called the hippocampus, in which spatial information is processed. There is evidence that in voles and kangaroo rats males do have larger ranges, better spatial ability, and a larger hippocampus than females (Gaulin and Fitzgerald 1990, Jacobs and Spencer 1994). Although evolutionary psychologists have extrapolated from these findings to humans, to do so is far from justified speculation.

Irrelevance of the Hunter–Gatherer Model for Sex Differences in Spatial Ability

The model of the hunter–gatherer tale causing differential brain development and function raises several objections. The nine points we are going to outline here should demonstrate that we are not convinced of the merits of this model.

First: we have little detailed knowledge of what human activities were really like during the four million years of human evolution—a rather remarkable time span that is now explained away in one arcadian model of rural life, complete with home (for women and children) and the world of work farther afield (for men). We cannot tell from the limited remains available to us today how prehistoric human beings really spent their days, how they divided up time according to sex, and whether human groups across time and space actually followed the same patterns. Indeed, we may never know or never know enough! All these assumptions may be no more than retrospective projections from known sexual divisions of labor in the recorded past and present.

Second: there are theories of human prehistory speculating that, for very long stretches of time, humans were actually not hunters but scavengers, and hence men did not go about killing animals (the hero coming home from dangerous pursuits may be a myth of the nineteenth century).

Third: another theory suggests that *Homo sapiens* was a successful hunter only because dogs joined in and did the hunting for them, in which case neither males nor females would have required differentiated spatial ability (Taçon and Pardoe 2002). In other words, and to use the jargon of evolutionary psychology, the Environment of Evolutionary Adaptedness (EEA) did not exist.

Fourth: the increasing dispersal of the human species over the globe has also suggested that a considerable part of prehistoric human activity was largely nomadic or semi-nomadic. Hence, a whole group/family/clan or tribe would move together—children, women, men, and the old members alike. Nomadic existence requires mobility of the entire group; it is difficult to see how joint activity of this kind would make a strong case for sex differences in human adaptations. As Mandler (1997) rightly said, presumptions of uniform or fixed behavior pattern may not have existed and this may "require us to think again about primitive reconstructions of imagined behavior of our ancestors" (p. 111).

Fifth: the idea of the hunter–gatherer society is not a unitary concept although it is often described as such. In fact, human development across the globe has been extremely uneven and has followed different paths according to climate (e.g., Ice Age effects), region, culture, and degree of exposure to other human groups. Those groups that were isolated from other groups for very long periods of time because of their inaccessibility have tended to acquire a state of equilibrium and discrete social structures. That pluralism is seen today, as we have highly urbanized societies at the same time as we have nomadic, semi-nomadic, hunter–gatherer peoples, and farmers with ancient farming techniques. Also, the so-called time of the stone-age man is largely a reference to European prehistory. Later, there existed more advanced societies with sophisticated cultures and technologies at the same time as humans in Europe eked out an existence in small camps or caves (Bloom 2000). By implication, any adaptations of these groups should differ somewhat.

Sixth: from the relatively scanty records we have of prehistoric human life, we cannot be confident in claiming that we know without a shadow of a doubt that all prehistoric societies over long time spans of tens or hundreds of thousands of years were socially organized in the same way; it is unlikely that all fitted the model of women scratching around in the dirt to find food and men braving the world by hunting. We can only say that there are existing tribes and societies today that send the women out while the men stay in the compound (Barnard 1999). In some New Guinea groups, men wear the adornments and women actively provide all material goods for the group. Ernst Bloch (1985) coined a very important historical concept that roughly translates into English as "the simultaneity of the non-simultaneous" (*Das Gleichzeitige des Ungleichzeitigen*), meaning that the history, including the prehistoric past, of humans is not one of uniform development. It would be difficult, therefore, to pinpoint when evolution of the brain stopped, who had the hunter–gatherer "genes" firmly implanted in their genome, and by when.

Seventh: large time spans have elapsed in which many human societies have engaged in basic agriculture. Eleanor Leacock, an anthropologist, took sociobiologists to task over the confusion between the nomadic and sedentary lifestyles as farmers. In hunting and nomadic life (i.e., during the period when human brain development stopped (allegedly)), she found little evidence of aggression and fighting. However, later, sedentary lifestyle led to violence (see Casey 1991). Over thousands of years, and perhaps tens of thousands of years, men were not hunters at all but shepherds, sitting at the entrance of makeshift huts or tents and surveying the flocks, passing the time making music or carving. The Old Testament is one of the first written sources about this lifestyle, and one can find many examples of shepherd cultures of men until this day. None of these activities would seem to be strong candidates for retaining genes that determine the cognitive abilities attributed to men by evolutionary psychologists. It is, we believe, obvious that they have failed to grasp the variety and fluidity of human existence.

Eighth: Moir and Jessel assume that males in prehistoric times had more stamina and were stronger than females. There is no evidence to support this view. Indeed, humans only a few centuries ago were a good deal smaller. Women in many agricultural societies both in the past and today (not just a few but millions of them) have accepted that their lot is that of hard labor. In many subsistence farming contexts anywhere in the world today, women do the physically hard work. Even in a modern country like Greece, it was still common rural practice in the 1950s for a future mother-in-law to test a potential bride for her physical strength and heavy load-carrying ability before she consented to her son's choice in marriage (Kaplan 1992).

From the armchair of a university office in an advanced Western society, semi-romantic myths of males as the ones with stamina and strength may seem plausible, but we would like to invite these theorists to go into any developing country right now and see for themselves who carts the wood, who carries the water, who plows the fields, and who makes the new roads. It is an extraordinary arrogance to disregard such current evidence and belittle the hard labor of women for the sake of a convenient model of explanation of genetic differences. As a United Nations statement said, on the basis of very well researched empirical evidence, women of the world do 90 percent of the world's work and own 10 percent of the resources. It is also a myth that women all over the world stay at home with the children. In fact, most women in rural areas, even in Western industrialized countries today (Kaplan 2000a), working on the farms, in the fields, and with the animals, have to take their children with them, often carrying them for long stretches of the day. Who has to have stamina?

The idea of differentials in stamina invites some theoretical counterproposals. One could argue that the stamina required by women is so great that, if we play devil's advocate and apply the theoretical constructs of some sociobiologists and evolutionary psychologists, we would expect some genetic adaptation of females for a potentially better performance in strength and, to use their line of argument, cognition also. Moreover, we could turn the assumption around and say that it is likely that the rise of capitalism, processes of urbanization, the development of a middle class and concomitant lifestyle changes, accelerated by technological inventions, might have led to remarkable physical changes in women and men in the western world, such as (1) a decline in physical stamina of those women who moved away from the fields and into drawing-room parlors, while men retained physical sports, with concomitant development of stronger sexual dimorphisms in physical appearance between men and women, (2) gaining in height (for both men and women due to better nutrition), and (3) increased leisure time for middle- and upper-class women, affording the opportunity to engage in activities of the mind. There is some proof for all of these claims, but we would not wish to build grand theories on the basis of these assumptions and would draw attention to some of the pitfalls of this kind of thinking.

Finally, a point that Howard Bloom (2001) raised recently is that implicitly, evolutionary psychology seems to have accepted the concept that evolution is always slow, or, as he put it, there is a "speed limit of genes" (p. 205) requiring thousands and even millions of years for any substantial changes to occur and, further, that these changes occur at an orderly and predictable pace. In fact, adaptations can occur in just one generation if the selective pressures are fierce enough (e.g., in times of drought or sudden change of climate) and they may involve DNA in the mitochondria. To repeat here the often cited example of a species of finch observed in the Galapagos Islands that adapted its beak size by selection in just one generation: when a drought had withered all fruits normally eaten by that species, most birds perished from starvation. The only survivors were those with a beak strong enough to crack open a particular nut for sustenance. The survivors produced offspring with strong beaks and the species changed to this new phenotype. This research is in its infancy, but there are strong doubts being raised about any assumed nature, speed, or degree of adaptation and whether it involves only the genes (DNA in the cell nucleus) (Bower 1999, Nachman et al. 1996).

Even if we were to accept the truncated and extremely simplified account of prehistoric human existence put forward by evolutionary psychologists, we might argue the converse, using their own frameworks and assumptions: that women required better spatial abilities than men because they foraged for food growing in various localities (Eals and Silverman 1994). This form of search would be enhanced by an excellent spatial memory of the location of objects placed in arrays (patterns); there is evidence that women perform better than men on tasks designed to test this type of spatial ability (McBurney et al. 1997).

Spatial ability is not a singular concept; it can be expressed in different ways. Recognition of this fact is of major importance in refuting any simplistic claims of a positive association between males and spatial ability. It is just as important to consider the potential role of learning different forms of spatial ability that might occur quite subtly during early life. To place this genetic assumption on the footing of human development into prehistoric times further diminishes the argument that any observed differences in spatial ability between males and females have their roots (causes) in genetic adaptation.

LANGUAGE AND MATHEMATICS DEBATE

Interpretation of exactly what specific cognitive tasks measure and how that relates to behavior in everyday life is difficult, if not impossible. The complexity of specific cognitive tasks is highlighted by the fact that tasks we might have thought of as being at least somewhat related, such as spatial and mathematical tasks, have been shown to have rather little to do with each other and may be entirely independent skills (Friedman 1995). However, as might be expected, verbal scores on one task are much better predictors of verbal scores on another task than they are of scores on rotation tasks (Gallagher 1989). For whatever reason, there may be sex differences in the amount of carryover of performance on one task to another; for example, the prediction of performance on a mathematics test based on performance in a rotation task has been found to be stronger for females than for males (Casey et al. 1995).

Geschwind and Galaburda (1985) postulated that sex differences in behavior are caused by the action of testosterone, which retards the development of the left hemisphere. This, they claim, explains why men have superior mathematical abilities. Although no aspect of this hypothesis has ever been tested in humans, the views of Geschwind and Galaburda have been widely accepted and propagated in both the popular and scientific media. In the 1980s, it was repeatedly and confidently announced that there is now no longer any doubt that boys are superior in mathematical and spatial ability and that this has a biological basis.

From the 1950s through the 1970s there were obvious social pressures against girls performing well in mathematics. For example girls who performed well in mathematics and science were said to be less attractive to boys than were other girls (Harper and Heath 1972, Karkau

1976). Often, strong social sanctions were applied if scientific interest and ability were detected in a girl, and the gender schema of the parent generation discouraged achievement (Jacobs and Eccles 1992, Jussim and Eccles 1992). Not surprisingly, at university, attendance of women in "hard core" science courses was very low well into the 1980s (Barnes 1988). Before the women's movement lent some support to the validity of scientific interest among girls, it was not uncommon for male fellow students and teaching staff in science to be scathingly derogatory to female students, with overt attempts to embarrass them publicly. It was this treatment, not as a singular event but throughout the entire period of their education, that made the study of physics or mathematics often difficult for female students, becoming more negative as the years progressed (Chipman and Wilson 1985). A number of studies in the 1980s showed that these social factors are the main source of sex differences in mathematics performance (Barnes 1988, Chipman and Wilson 1985). It cannot be surprising to find that social discrimination carried over into cognitive tasks involving mathematics and other skills perceived to be related to mathematics or science (Rech 1996).

Almost always, tests to establish aptitude have been measured using simple tasks and then extrapolated as if the performance score applied to a far wider social arena. For example, mathematical ability is usually measured in a pen-and-paper performance task given once only to each subject in the study. Past experience, anxiety about the material being presented in the task, interaction between tester and subject, and many other variables are ignored so that a single numerical score can be given to each subject, and this then can be incorporated into statistical analysis for sex differences. The results of such testing are then generalized to a much wider context of so-called mathematical ability including simple addition of figures, computer programming, applied mathematics, pure mathematics, and other related skills. This is not to say that we are trying to rule out the idea that sex differences in brain functioning may exist. Rather, it is important to place current findings into some context (impinging variables) so that genetic/hormonal causation is not simply assumed.

We are not saying, therefore, that sexual differences in brain function do not exist. Men and women might, for example, use different regions of their brain to achieve equivalent results. Richard Haier and Camilla Benbow (1995) made brain scans of males and females while they were solving mathematical problems and obtained results showing sex differences in the regions of the brain that were active at the time. Before the experiment was carried out, each subject was categorized as either a high or an average performer on the basis of scholastic aptitude testing in mathematics. Immediately after this, a positron emission tomography (PET) scan of the subject's brain was carried out. A sex difference was observed in the high-scoring group: the men had more activity in their temporal lobes of the brain than the women. Other areas of the brain were also active but these were not found to differ in the women and men. The average-scoring men did not have this higher level of activity in the temporal lobes, nor did the average-scoring women. High-scoring women and men (referred to as gifted) were therefore using different regions of their brains to perform the same mathematical tasks. This might suggest that they go about solving problems using different strategies. Note that it is the gifted males who are the exception, since the pattern of activity in gifted females did not differ from that of average females. It is only a small step from here to say that the gifted males were not only exceptional but superior. The study provides no evidence to explain why gifted females differ from the other females, and this is a complicating factor.

Moreover, there is another explanation for the result obtained. We cannot tell whether the males were actually using their temporal lobes for processing the given problems, since it is possible to think about more than one thing at a time and so another function of the brain could be going on in the temporal lobes. Also, no analysis was made of levels of anxiety about the task or about mathematics in general. The sexes could well have differed in this factor. Such confounding inputs to the results could have been established just after the subjects had been tested by asking them some simple questions or by measuring heart rate or other stress responses during the test itself, but this was not done. To do so would require a broader view of possible factors influencing the result. We know from animal studies that exploratory behavior is impeded by stress or anxiety. In other words, the evidence is not conclusive. All the results tell us is that there is something different about the brain activity of the "gifted" men and women tested but what it is and what causes it remains unknown.

Even though Haier and Benbow (1995) found that there was a mild but statistically significant association between the amount of activity in the temporal lobes in the male subjects and the number of mathematical problems solved correctly during the experiment, it still does not prove that the temporal lobes are the region where the problem solving is taking place. It is possible that activity in the temporal lobes reflects more motivation to perform well on the task, and so reflects motivation rather than problem-solving ability. We just cannot tell what is the explanation until researchers carry out all the additional tests that are needed before drawing conclusions.

Sex differences observed with imaging of brain activities have also been found by Bennett Shaywitz and colleagues (1995). They used functional magnetic resonance imaging (fMRI) to examine the patterns of activity in male and female subjects' brains while they performed a number of tasks using words (see also Pugh et al. 1996). One of these tasks revealed a sex difference in the images obtained. When the male subjects were tested by asking them to say whether or not two strings of nonsense words rhymed, nerve cell activity was higher in a region in the left hemisphere, called the left inferior frontal gyrus. When the female subjects were asked the same questions, the activity was higher in both the left and the right inferior frontal gyrus than in other parts of the brain.

Although this apparent sex difference in the fMRI images could be criticized on the same grounds as above, it is worthwhile to examine this further. A similar bias to the left hemisphere in men and not in women had been identified by some previous research (summarized in Bradshaw 1989, Bradshaw and Rogers 1993). It indicates that women use regions of both the left and the right hemispheres when processing language, whereas men use the left hemisphere more than the right. This was found by flashing words on screens located to the left and right of the subject. The subject had to look straight ahead so that the words were visible at the edge of the left or right field of vision, thus ensuring that the left hemisphere processes the words seen in the right field of vision and the right hemisphere processes the words seen in the left field of vision. The subject was asked to read the words. Performance was better when the words were in the right visual field compared to the left visual field. This was the case for both women and men, but a comparison of performance when the words were presented in the left visual field showed that the women performed better than the men. This result indicates that women might process some aspects of language in the right hemisphere as well as in the left but, again, we would need to control for motivation and anxiety before we could be sure that the difference is specifically related to processing of words and not to something general and nonspecific. Added to this, it could be a difference influenced by culturally biased experience.

In addition to this sex difference in use of the left and right sides of the brain for language processing, there may be a sex difference in the use of the frontal and posterior regions of the brain hemispheres. Doreen Kimura (1983) worked with subjects who were unable to produce speech (known as aphasia) following brain damage caused either by a stroke or a brain tumor (cf. Kimura 1992). She determined the location of the damage using electroencephalographic recording (EEG), which involves placing electrodes on the scalp in various places and using them to measure electrical activity in the brain. This is an older, less expensive, and less accurate technique than fMRI or PET scanning. Within the limits of accuracy of this method, she found a sex difference: the women suffering from aphasia were more likely than the men to have damage in the frontal region of the brain. In men suffering from aphasia, the damage was more likely to be in the posterior region of the hemispheres. These results could, however, have been compromised by the fact that more of the women used in the study had tumors than did the men, and more of the men had damage caused by a stroke.

A subsequent study by Andrew Kertesz and Thomas Benke (1989) examined only stroke patients with aphasia and found no sex difference in the location of the site of damage in the front or back regions of the hemispheres. In their study, the location of the damage was found using computer imaging and was therefore considered to be more accurate than the method used by Kimura. This result clouds the issue of whether there is a back–front difference in where women and men process language in the brain. Relevant to our discussion above, Kertesz and Benke also found no evidence of a sex difference in processing of language in the left versus the right hemisphere.

Whether or not women and men do use some different regions of the brain to produce speech or process language (and further research is needed to decide this), merely finding this difference would tell us nothing about what *causes* it. There would be no basis to conclude that cognitive differences between the sexes stem from differing hormonal influences on brain development, as did Kimura (1992). Before we could arrive at this conclusion we would have to investigate just what factors might affect the sex difference. Despite the certainty with which this particular claim is phrased, there is no strong evidence on which to base it. Indeed, such evidence would be difficult, if not impossible, to obtain since it is impossible to tease apart any separate contributions that biology and culture might make.

Another study using PET scanning reported differences in the pattern of activity in male and female brains at rest, meaning that the subjects were not asked to perform any specific mental task while their brains were scanned (Gur et al. 1995). The amount of activity in the temporal limbic system was higher in males than in females. (The temporal limbic system refers to a number of regions inside the temporal lobes; it is one of the parts of the brain involved in the expression of emotions.) The activity in the cingulate gyrus was higher in the females than males. This is another region of the limbic system, also linked to emotional responses but perhaps at a higher level of processing. The cingulate gyrus is also associated with perception of pain and is involved in the relay of higher mental processes of the cortex to lower regions in the hypothalamus. Gur and colleagues interpreted their results as indicating that sex differences in cognitive and emotional processing have a biological substrate, but this conclusion was not warranted by their results. For any number of reasons, women and men may think about different things when they are "at rest" during the experiment; these thinking patterns themselves could explain the different patterns of brain activity. This might depend not so much on genetic causes but on differences in past experience and differing immediate reactions to the experimental procedures.

Many of these studies have used large numbers of subjects, since only then do some of the sex differences emerge. Then it has to be recognized that there is a great deal of overlap between the scores for females and males, and that this overlap might be as important, if not more so, than the small difference between the groups. But scientists and others have a strong tendency to focus on differences. This is one reason why similarities of male and female mental abilities are overlooked or, at least, not emphasized. In fact, this has happened to such an extent that some writers have seen it timely to draw attention to similarities rather than differences (e.g., Connell 1987, Riger 1992). Nevertheless, it is not unreasonable to question how the differences, even though smaller than often portrayed, might have come about.

PART THREE

In Perpetuity

When nature formed you, she doubted for a moment
Whether to offer you as a girl or a boy,
While she sets her mind's eye to settling this,
Behold! You come forth, born as a vision for all.

Hilary the Englishman
(twelfth-century English poet)

7

Designated Sex Development

GODS, GODDESSES, AND MALE AND FEMALE COMPETITION

The designation of who is female and who is male is inextricably linked to a long-raging battle over the relative evolutionary importance of female and male. Formerly, the battle was fought in cults and religious beliefs. Today, in a secularized world, it has been shifted largely to science and to molecular levels. Broadly, social organization and relative role accommodation of the sexes has tended to be in tune with beliefs of the role of the egg and the sperm. In prehistoric times, women were often placed on pedestals as symbols of fertility. Matriarchies tended to be associated with the belief that the entire role of reproduction rested with women. The discovery of the sperm as an important contributor to reproduction wrested power away from women but not immediately and not entirely.

Throughout many centuries, goddesses of fertility were revered around the world. Those in Asia Minor and Middle Eastern countries eventually found their way to Europe (first Greece and then Rome). There were male gods and female gods side by side, but goddesses of fertility were central to many spiritual beliefs. For instance, the legend of Attis tells of a young shepherd much beloved by the goddess Cybele, who was not just any goddess but the mother of all gods, and was said to have her home in Phrygia. The various remnants of this legend are particularly telling as they demonstrate the shift from adulation of female goddesses to male gods. Attis, in some versions, was thought to be Cybele's son, in others, her lover. Attis was either the son of Cybele or the son of Nana, a virgin who had conceived him by putting a pomegranate or a ripe almond in her bosom. In Phrygian thought the almond represented maleness because the blossoms of the almond were the first to bloom in spring. Attis eventually died. There were two accounts of his death, one in which he was killed by a boar (as Adonis) and the other, that he unmanned himself under a pine tree as an ultimate sacrifice to Cybele, whereupon he is said to have turned into a pine tree. All male priests serving Cybele were castrated. The temple ceremonies for castration were on the Day of Blood, when men castrated themselves with a sword, then held their severed parts and danced with them, eventually passing them on to select persons or households who received them as sacred relics. The castrated males were then put into female clothes and wore them as a sign of honor.

Romans adopted the worship of the Phrygian mother at the end of the third century BC, but over time, there was another twist to the story. Attis, the beloved god, was resurrected. On March 25, then considered the vernal equinox, the divine resurrection occurred and was celebrated as a Roman carnival called the Festival of Joy (*Hilaria*) (see Frazer 1974). The resurrection of the male became the true cause for celebration. Aspects of the legend, such as Attis's birth to a virgin, his death due to self-sacrifice, and his resurrection, all well known in the first century BC, invite comparisons to aspects of the Christ story, with its new emphasis on father and son.

Once the Judeo-Christian tradition triumphed over Rome and the last vestiges of the ancient world were buried, any semblance of earlier worship of goddesses had gone. Mary, the vessel for another virgin birth, was the only female representation left at the start of the modern Western Christian era. Importantly, the Judeo-Christian tradition was instrumental in raising sperm to the central substance for life on earth, a divine ingredient that could not be spilled or wasted. Hence, masturbation by males was forbidden and came under very serious surveillance from the early eighteenth century on (Foucault 1977), and homosexuality had to be punished by death. Most death sentences and forms of degrading public punishment were lifted only in the nineteenth century (West 1977). Women such as midwives, who knew too much about contraception, had to be burnt as witches and the Inquisition guarded the male grail of life and divinity (Ehrenreich and English 1976, Hammer 1977, Trevor-Roper 1969).

As societies became more secular and as science advanced and it became clear that egg and sperm were needed for reproduction, the role of women could be revisited on social grounds (see Martin 1991). The competitive idea of the importance of male or female in a reproductive sense has not become merely a vestige of the past that modern thinkers and societies have left behind. Instead, these ideas can be found today in many areas of science, played out in methodologies that often reinforce them.

A recent example of such male–female competitive statements was engendered when Wen-Hsiung Li and Kateryna Makova (2002) published an article on gene mutation with the emotive title "Strong Male-Driven Evolution of DNA Sequences in Humans and Apes." They reported results based on study of the male Y chromosome and arrived at estimates that males produce over five times more genetic mutations (the raw material for evolution) than do females.

Their findings created excitement for two reasons. First, they refuted earlier calculations by the Human Genome Sequence Consortium in 2001 that mutation rates were only about one to two times higher in males than in females. The new results were far more dramatic and were advertised as a male win, as exemplified by the heading "Men Regain Evolutionary Driver's Seat" on a press release of April 10, 2002, from the University of Chicago. What makes this type of announcement so remarkable is the partiality for an apparently "winning" party, the male of the species—even at the chromosomal level. Males are said to drive human evolution. Egg cells divide 24 times, with most of these cell divisions occurring before birth of the female, while sperm stem cells divide throughout life and accumulate mutations. Secondly, the mutations are said to be caused mainly by random errors during cell division rather than by environmental factors. Hence, the conclusion that genes alone (without environmental influences) drive evolution. The entire equation could then be read as follows: genes drive evolution via mutations that are speeded up when there is a Y chromosome. X chromosomes play only a marginal role. Hence, men drive evolution and behavior. There we have it: Eve as a rib of man, he as the creator, and she merely as an evolutionary by-product, the vessel through which his genetic mutations can operate.

Mutations we need to stress are *errors* occurring when the DNA replicates. It would have been easy to argue that males seem to be the main source of genetic problems since their sperm cells continue to accumulate errors throughout life. In response to this suggestion the scientists pointed out that the mutation rate is very low and affects all of the DNA, 90 percent of it being noncoding (i.e., not expressed in proteins in the cell). Thus, in their opinion, the male genome is both driving evolution and exonerated from responsibility for generating undesirable mutations.

This celebrated importance of the male in the evolution of the human species comes at a time when women seem to be moving back into traditional thinking in society. There are now books on how to be a lady (Cleary and Von Mueffling 2001), and books that show how harmful and unjust we have become to boys (Sommers 2000). In this climate of opinion, exaggerated claims of male superiority have enormous implications in terms of social attitudes. The point is not so much whether Li and Markova were correct in these new findings but how these findings are translated into the social, political, and cultural arenas. To understand the functioning of a complex process of cell division in reproduction, whether from the point of view of the embryologist or the geneticist, is a proper scientific goal. To draw *social* conclusions from cell divisions (as the press release overtly did) is absurd. Yet so steeped is our thinking (including that of some scientists) in polarized and gendered dichotomies that the research reporting seems to be driven as much by a desire to validate the ingrained "rightness" of gender hierarchies (domination versus subjugation) as it is by the sense of discovery of the detailed biological mechanisms of gene mutation.

POLARITIES AND SEXUAL DIMORPHISM

Perception of inflated, even false, polarity starts by assuming that maleness and femaleness are entirely opposites, even at the cellular level. Questions of sameness, degrees of differentiation, or certain role similarities are rarely seen as newsworthy or worth being considered seriously. Ironically, we live in an age in which multiplicity, diversity, and chaos theory have suggested that simple binaries or linear relationships may have little explanatory power but too much political currency. Yet current findings related to sex or gender differences are often couched in terms that reinforce stereotypical male and female processes. This is evident even in discussions at the cellular level (Fausto-Sterling 2001).

Moreover, this kind of male–female polarized sensationalism is at odds with the diversity of biology. The human Y chromosome, so Page and co-workers (1987) found, contains the testis-determining factor gene, and thus any human male needs one Y chromosome to differentiate as a male. Yet there are partially phenotypic males with XX chromosomes (when the adrenals produce excess androgens) and phenotypic females with XY chromosomal structure (as in testicular feminizing syndrome, in which the cells of the body fail to respond to testosterone). Whether such individuals are considered to be male or female depends on the morphology of their external genitalia (cf. Fausto-Sterling 2001), although the medical profession often ignores the phenotype and refers to "genetic males" and "genetic females." An added complexity occurs when a sequence of genes that is normally located on the Y chromosome relocates to an X chromosome. Furthermore, it is possible to have external genitalia that are not fully differentiated into one or the other sex. There are many other subtle intermediary steps at morphological and hormonal levels.

It is sometimes difficult to decide whether the baby at birth has a penis or an enlarged clitoris. In the past, mistakes in assignment have occurred and have only been uncovered when the growth of the child allowed clearer differentiation. In girls this condition is called adrenal hyperplasia (enlarged clitoris often mistaken for a penis). In these cases, the child often opts to stay with the initial designation of sex/gender, even though both physiology and anatomy may indicate otherwise in later life. Some hospitals wait for the results of genetic tests before they announce the sex of the newly born child. But despite efforts in sex assignment, errors have occurred when attempting to match hormonal and chromosomal sex. There may have been a newborn, deemed to be really a male but lacking a penis, who had the external structures surgically corrected, and the resultant male later found that she identified as a female. In other words, the medical and biological conclusions, based on hormonal, genetic, and external genital reference points, were still not sufficient to accurately predict the individual's sexual identity later in life.

The reason such predictions can be so inaccurate despite genetic and anatomical information is that the decisions are based on the anatomy of the genitalia. From anatomical division into the male and female genital sex this information is extrapolated to the brain and to a functional division between femaleness and maleness. But differentiation of the brain, the most complex organ of the body, is not the same as for the genitalia, as Beach (1971) pointed out long ago. It seems that few were ready to listen to Beach then. As an example, in male rats, nerve circuits for female typical sexual behavior are present: the male animal can be induced by appropriate stimulation on hormonal administration to perform female sexual behavior and other patterns of behavior more characteristic of females (Nordeen and Yahr 1982, Sodersten 1976). Hence, the development choice of the brain is not a discrete one of either male or female, even in rats. It is also certainly the case that human males and females can perform both male- and female-characterized types of sexual behavior, assuming that it is even possible to make such a division between these types, since human sexual practice is malleable and varied.

Apart from phenotype and genotype, gender identity is a further distinct category in the biomedical field. The gender identity of a person is considered to be the sex that she or he feels herself or himself to be, irrespective of biological sex. An individual may, for example, feel that she is a man even though her physique and aspects of her biology are typical of a woman. Or a person who is physically a man may feel that he is really a woman. In such cases we speak of transsexuals or, more recently, transgenders. The sex versus gender issue here is completely entangled, as biological sex and gender identity are not consistent with each other. But the gender identity felt by an individual may not be expressed in that individual's general behavior. It is possible for a person with a woman's biology (sex) to have a male gender identity and still behave in a feminine way (i.e., express a female gender role).

We have outlined some of the overlaps in sexual designation of individuals. Skeptics may say that we have dredged up some extraordinary and marginal cases just to force a view that similarities are important when, in fact, they should be regarded as rare and even as abnormalities. Sexual differentiation of the genitalia is said to be one of the clearest biological differentiations even before birth, genetically coded and evolutionarily essential, and differences from the normal patterns are aberrations that should not be regarded as worthy of making a general theoretical point. But variability in the differentiation of physical sex is a good deal more common than believed. There are sufficient biological reasons that made it acceptable and important to deal with variability rather than fixed polarities. Viewing the world in models of polarity ignores continuities, overlaps, and gradations that occur in the physical expression of sex differences. Add to this the differentiation of behavioral sex and the picture is completely blurred.

SEX DIFFERENTIATION AND DEVELOPMENT

During fetal development, undifferentiated tissue differentiates into either a female or male genital structure. The ovaries or testes begin to secrete their hormones very early in development, even before birth, and it is now possible to determine the sex of the fetus, using ultrasound, with a reasonable degree of accuracy. In fact, it has become common in Western countries to assign a sex to an expected child. One wonders what effect this early assigning has on the behavioral development of the fetus and child. Generally, however, a child is assigned to either the male or female sex immediately after birth or soon thereafter, if genetic typing is employed. Once such an assignment has been made, there are expectations of a "normal" course of development to sexual maturity appropriate to the child's assigned sex. These are quite rigid assumptions, despite substantial variations at the level of physiology and onset of sexual development, as discussed previously, and despite the known effects of environmental factors such as nutrition and stress, particularly during crucial stages, on the sequence of events and their outcomes. In general, the pattern of development matches the genetic sex but the numerous factors that can influence sexual development might be diversifying rather than channeling.

June Reinisch and colleagues (1991) see the interaction of biological and environmental factors in terms of a "multiplier effect." In their view, genetic and hormonal differences cause sexual differentiation of the genitalia and the brain prior to birth (discussed earlier) and lead to behavioral differences between the sexes at birth. As they see it, from birth on, these behavioral differences between the sexes are augmented by successive interactions between the individual and the social environment. At puberty the levels of the sex hormones secreted increase greatly. At the same time, when the physical differentiation of the sexes increases, social expectations and social interactions, in their view, magnify the behavioral differences between the sexes, thus enhancing sex differences in behavior. We would suggest an alternative view that the diversity of factors impinging on sexual differentiation broadens the potential for both physical and behavioral expression but that this may be blunted and channeled by social norms, thus constricting behavior to sex stereotypes. Some of the factors impinging on the individual are thought of as being internal (not only genes and hormones but also the child's thoughts about himself or herself) and others external (inputs from the environment, particularly social attitudes) to the individual, and there is no clear separation of internal and external factors; they interact.

It is common to think of the individual organism, be that a human, an animal or even a plant, as beginning life with some sort of basic program encoded in its genes (Gould and Marler 1987) and that learning (or experience) modifies this program as the organism matures. The program is seen as being within the organism and its modification by experience comes from outside the organism. In this way we conceive of some sort of action and reaction between inside and outside the organism, but this is not really an appropriate description of the developmental process. The inside and outside mutually interact with each other. The entire process of an individual's development is not merely a battle between nature (genes/inside) and nurture (experience or learning/outside) but rather a dynamic interweaving of processes within a system that is inseparably the organism and its environment. Genes are neither central nor primary to ongoing developmental processes, and the same is true of the effects of experience. Thus, we might argue that there is the possibility of more flexibility in sexual development. This view contrasts to the ever tightening of the behavior of the individual to his or her genome, as proposed by Reinisch and colleagues (1991).

Our biology makes fewer distinctions between the sexes than does our social world. We know that biological factors contribute to the determination of the physical characteristics of sex, the construction of the genitalia, and a number of other physical traits. However, it is not a direct, invariant, or unbroken chain of causation from genotype (genetic) via several hormone levels to either the female or male phenotype (physical type). Biology does not make a discrete choice, and variation enters the system at all levels during the process of differentiation. This results in a broad range of morphological (structural), physiological, and behavioral characteristics in genetic males and females. As a partial understanding of this process, as vom Saal (1983) has stated, it is the very fact that sexual differentiation is affected by sex hormone levels that guarantees marked variation in phenotype.

SENSITIVE PERIODS OF DEVELOPMENT

During development there are sensitive periods when certain things are learned rapidly and with greater ease than at other times. The timing of these special stages in development can be programmed in the genes but exactly what is learned during each sensitive period is not part of the program. Experiments show that sensitive periods may also affect brain development and sexual differentiation.

Humans are not born in an advanced stage of development. The human brain does develop before birth but much of its development continues after birth. It is rapid in the first months after birth but actually continues for several years. Given that, on average, boys secrete higher levels of testosterone than girls both before and after birth, it is possible that any effect of testosterone on development could happen before or after birth, and in the latter case the effects could be indirect via changing the parents' behavior, as we have discussed in Chapter 3.

Research in the 1970s indicated that exposure of female human fetuses to high levels of androgen masculinized their behavior in later life. We consider it important to discuss this research in some detail because it has been so influential in the whole field of the development of sex differences. The exposure of the fetuses to androgen resulted because their mothers had taken the drug progestin, to prevent miscarriage. At that time, many women were prescribed the drug. Only later was it found to have an androgenic action that masculinized the genitalia of some of the female children born to mothers who had taken it. John Money and Anke Ehrhardt (1967, 1972) decided to use this situation to test whether exposure to the drug might have masculinized the behavior of the girls. They chose progestin-exposed girls without obvious masculinization of the genitalia. The aim was to control for any potential effects that having masculinized genitalia might have on the way the girls were treated by their family and others. The study included an additional test group of girls with adrenal hyperplasia, a condition in which the adrenal gland produces higher than normal levels of testosterone, as well as other androgens and stress hormones. These subjects, exposed to higher levels of androgens, would have had masculinization of the genitalia because that is how the condition is detected, although this was never discussed by Money and Ehrhardt.

One procedure used by Money and Ehrhardt was to interview the girls' mothers by telephone and to ask them whether their daughters were tomboys or liked wearing girls' or boys' clothes, playing with girls' or boys' toys and so on. Another approach was to test some of the girls and so to assess their gender identity, attitudes to work and marriage, IQ scores, and whether they had homosexual fantasies. The results showed a pattern of non-traditional preference for tomboy play, boys' clothing, higher IQ scores, and choice of career over marriage. But most of the conclusions reached by Money and Ehrhardt were based on data collected by interviewing the girls' mothers, even though it is well known that information obtained by such retrospective reporting can be unreliable, particularly when given over the telephone. These results were compromised by inadequate controls: some of the sisters of the test subjects had the same patterns of behavior even though they had not been exposed to progestin or high levels of androgens.

A follow-up study of girls with adrenal hyperplasia by Susan Baker and Anke Ehrhardt (1974) was better controlled and showed that their only difference from control girls was in playing more energetically. This play was now referred to as "increased energetic play," rather than the value-loaded "tomboyishness." A more recent study by Sheri Berenbaum and Melissa Hines (1992) found that girls with adrenal hyperplasia tested at ages from 3 to 8 years preferred to play with boys' toys, whereas no such preference was expressed by their female relatives without adrenal hyperplasia. In all of these studies, however, the parents of the androgen-exposed girls may have contributed to the way in which their daughters behaved, because it is likely that they reacted to the masculinization of their genitalia. In fact, a study by Slijper (1984) found that parents of girls with adrenal hyperplasia exaggerated their reports on the preference of their daughters for energetic play, possibly because they harbored doubts about the sex of the child. These doubts expressed by the parents could be transferred to the child, who may respond by showing behaviors that are not so typical of girls.

Slijper also found that the main cause of increased energetic play (or "romping behavior," as she called it) and other apparently "masculinized" behaviors was the experience of being ill in early life, rather than being exposed to abnormally high levels of androgens. Children with adrenal hyperplasia fall ill because of excess loss of salt from their bodies; they are also hospitalized for corrective surgery of the genitalia. Slijper compared girls and boys with adrenal hyperplasia with girls and boys who had been ill with diabetes. Diabetes is not a condition that causes abnormal levels of the sex hormones but diabetic children have in common with hyperplasic children prolonged illness and hospitalization. Compared with girls who had no known illness (the control group), both the diabetic and the hyperplasic girls were more "boyish." The opposite trend was found in the boys: both diabetic and hyperplasic boys were more "girlish" than the control group of boys. So illness in early life alone can lead to gender atypical play behavior.

Other studies have followed up the suggestion that the androgen-exposed girls might have lesbian tendencies and found increased reported homosexual fantasizing but no evidence of increased homosexuality (Dittmann et al. 1992). It is well known that fantasies are not reliable indicators of sexual orientation as practiced. Despite the weak to nonexistent evidence for lesbianism, a number of influential researchers have continued to claim that lesbian behavior could be caused by exposure of girls to high levels of androgen before birth and that male homosexuality could be caused by insufficient exposure to androgens before birth. What is more, because stress suffered by mothers during pregnancy can lower the level of testosterone circulating in the blood of the fetus (Ward and Weisz 1980), Dörner and colleagues (1980) reasoned that maternal stress might cause homosexuality of the male child by reducing the amount of testosterone to which the fetus is exposed. A finding of increased incidence of homosexuality in men born during the Second World War seemed to support this hypothesis (Dörner et al. 1983), but the recorded incidence of homosexuality is a matter of disclosure. Reporting can vary with changing social attitudes and so may not reflect biological contributions. Fluctuation of hormones is natural and occurs in all individuals. In females, levels of the sex hormones fluctuate over the menstrual cycle and they change in both males and females with aging and in response to inputs from the environment, including stress. However, there is no good evidence that an unusual hormonal event is required for the development of homosexual identity and/or homosexual practice.

It needs to be said that other broad stages of development of the individual are fluid. In terms of the development of identities of the individual (male, female, heterosexual, homosexual, etc.), there are important variances; stages may have no abrupt beginnings or ends. Indeed, many descriptors of developmental stages are merely heuristic devices, as Weinberg (1984) pointed out, used to organize reflections and data. Some of the biological models of development, he noted, merely transfer notions of normality to individuals and further encourage a linear thinking of uniform developmental rates. These points are very important to make, particularly in appreciation of identity formation and changes in such identity that may occur over a life span.

FEMININITY, MASCULINITY, AND THE FAMILY

At birth the assignment of an infant to male or female is constrained by social views of what is female and what is male. This assignment has many ramifications for the parent–child interaction. In the past it related strongly to inheritance of the family's wealth, and it still does so in some classes and circles today. It also relates to career choice and social success, carrying on the family name, and so on. Knowingly or unknowingly, we teach female and male children different things almost from the moment they are born. We have discussed this in Chapter 3. Even mothers who are determined to treat their male and female children in the same way display different responses to girls and boys.

The cultural influences on the way we react to girls and boys are very persistent. Social differences are often considered to result "naturally" from biological differences, but biological differences may result from the influence of being raised in different social environments. Different environments may start, for example, by dressing baby girls in pink and baby boys in blue and continue by mothers speaking to girls more than to boys and encouraging (or discouraging) girls and boys to play with different toys and to show different amounts of aggression.

If there is any area of research in which the nature/nurture debate has never subsided it is in gender role development of children and, like any nature/nurture debates in the past, it has been bogged down by these polarities. The question is whether or not the genetic and hormonal discrimination into male and female translates into sex-specific behaviors, likes and dislikes, choices of toys, clothes, activities, occupations, and thus permeates all behaviors, or whether the basic grid of genes and hormones just broadly defines each individual's reproductive function (Raag and Rackliff 1998).

In the deterministic view, environments are not simply imposed on girls and boys by the society around them but the child seeks to put herself or himself in an environment typical for a girl or a boy. It is said that girls seek to surround themselves by different environments than do boys and seek to learn different things than do boys because their genetic program drives them to do so. Having XX or XY genes is said to shape attention to certain things and ensures that the child will develop interests typical for his or her genetic (and hormonal) sex. Although this idea acknowledges the role of society and learning in determining sex differences in behavior, it also subsumes this concept within a framework of genetic determinism. In essence, this concept of development encapsulates "the environment" within the genetic program for development. Genetic difference between the sexes is seen as the primary, if not supreme, factor guiding even those aspects of development that require learning. Any environmental influences on development are considered to be, ultimately, controlled by the genes.

The constructionist view, at the other end of the scale, holds that children are not just children. They are girls and boys with adults around them who have aspirations, hopes, and expectations that they impart to their children. Wishing the best for their children, parents may well encourage behaviors that will allow them to fit into society as girls and boys; this is also true of parents who try to treat all their children equally (Lytton and Romney 1991). Gender norms are applied very early, even among egalitarian-minded parents (Weiner and Wilson-Mitchell 1990). There is also plenty of evidence that sexual division of labor has retained currency, not just in the wider world but also in the home, and here male and female children get involved differentially and quite early (Hoffman and Hurst 1990). Most parents, even when both are engaged in full-time work, do not share the housework equally. According to gender studies in the United States, egalitarian males contribute about 30 percent to all household chores and both partners usually agree that this is fair (Lennon and Rosenfield 1994). An Australian study found that some chores, such as ironing, are never done by males, dishes rarely, and that the male contribution often starts and ends with some active participation in child care and with assisting in well-defined brief jobs, such as taking out the trash (Bittman 1991). Children usually observe these role differences in their parents and model their own behavior on them.

The question is whether this polarized approach to questions of gender performance and gender identity development will get us very far. Most studies that have been conducted to address questions of gender in this manner have a point to make. In terms of our scientific understanding, however, we are left in a position of either accepting or rejecting the various hypotheses. It is at these junctures where theories purporting to have solved all these difficult questions are so very tempting. Positivistic studies, even the best empirical studies, are limited by what they can study. They take a snapshot and a time slice and observe how people actually behave, how tasks are divided, how roles are developed, suppressed, or exhibited. Yet, families do not live together for the sake of dividing up roles, and the rational responses and comments that researchers might get from members of the family reveal very little of their inner life together.

Neither neo-Darwinian theories nor most empirical studies on sexual divisions of labor broach issues of love, dependency, and strong emotions such as jealousy, fear, despair, need for power, intimidation, or strong identifications, be this within a family or in society at large. At best we might be allowed to view this biologically as cooperation or competition in terms of "investment" of genes, or have it explained in terms of bargaining powers and power differentials. But the relationships within a group of people, some of whom are children, are very complicated by these factors; life in these microcosms can be heaven or hell on earth and is usually somewhere in between. The strengths and weaknesses of ties and fulfillment of emotional needs determine how well the individual will learn, how "tough" or ambitious the individual might be, and how successful she or he will be in the future.

There are predictors of success, and even health, but in the psychosexual domain, prediction is far more difficult. We cannot tell in advance whether children will be happy adolescents or adults, and we cannot easily predict their sexuality or the strength of their gender identity and how this will manifest itself socially in adolescence and adult life. We cannot predict what attitudes a particular child will have to his or her own body, how sexually active he or she will be, whether fulfillment of his or her sexual needs will be possible, or even if there will be sexual needs. We cannot easily do so because sexuality and psychology as a whole develop in a broad psychological and sociohistorical context; these worlds of the family and the disinterested society are linked by feelings, innuendoes, erotic tensions, and repressions.

Social norms tend to be explicit with respect to sexual conduct and, for the most part, tacitly assume and even enforce a heterosexual framework. They tend to be particularly explicit and even severe in relation to girls (Mead 1974, 1988). Moreover, sexuality, femaleness, and the body are bound together by the sex-exclusive role of child bearing and raising in ways that have given rise to extraordinary sets of rules governing the conduct of females over the ages. As far as we know, over many centuries at least, males have often liked to brag about their sexual exploits (true or imagined) while in many countries, females have been expected to refrain from sexual contacts altogether until marriage or have been presumed to have no sexuality of their own. In most generations of the past century, teenage girls have been taught to be shy about their sexuality or to postpone active sexual relationships to adulthood. There is an ominous silence about topics related to adolescent sexuality, except for some recent writing by young feminists themselves (Baumgardner and Richards 2000, Edut 1998). There has been relatively little said about adolescent girls in developmental psychology, and feminists in psychotherapy note that even their discussions have been relatively silent on adolescent girls, let alone include topics such as feelings and sexuality (Gilligan et al. 1991).

FEMININITY, MASCULINITY, AND CULTURE

One of the broadest cultural definitions of masculinity and femininity in the twentieth century was provided by a C. G. Jung follower, Erich Neumann (1954). He used the terms

> not as personal sex-linked characteristics, but as symbolic expressions. When we say masculine or feminine . . . this is a psychological statement that must not be reduced to biological or sociological terms. The symbolism of "masculine" and "feminine" is archetypical and therefore transpersonal; in the various cultures concerned, it is erroneously projected upon persons as though they carried its qualities. In reality every individual is a psychological hybrid. Even sexual symbolism cannot be derived from the person, because it is prior to the person. Conversely, it is one of the complications of individual psychology that in all cultures the integrity of the personality is violated when it is identified with either the masculine or the feminine side of the symbolic principle of opposites. [p. xxii]

Archetypal symbolisms of masculinity and femininity are claimed to be common to all cultures, and, if we assume this to be true, then there is no single way to describe sex traits in objective terms. Neumann, like Jung, understood the feminine to represent unconsciousness, darkness, and night, and the masculine as consciousness, light, and day. He used these categories historically and provided a sketch of human cultural development from darkness, meaning from matriarchy (note that the time of the worship of the egg and fertility coincides with darkness) to patriarchy and consciousness, as a period of light and day. The unconscious concerned psychic relatedness, and feelings or emotions, while the conscious represented logic, objectivity, and perfection. The Jungian contrasts of Eros (feminine) and Logos (masculine) add nothing that philosophers had not already identified in the eighteenth century, as did Rousseau, for instance.

However, despite the unflattering concept of the feminine as a symbol of a blind spirit of the night that does not know itself, Neumann did argue quite progressively that the translation of these symbols into individual psychology is not only a "complication" but also a *violation* of personality. Neumann is thus quite clear that gender symbolism is and ought to be situated at a very different level of explanation (high culture) than in individual psychology (everyday life). He is also quite clear that these symbolisms have no place in sociology or biology. Indeed, he is at pains to distance himself from an association with biology, as were Freud and Jung before him.

In other words, there is a strong tradition in twentieth-century psychology that argues firmly against the notion that femininity and masculinity are biologically determined, or even that they derive their meaning from observing men and women in everyday life. Culture, however, imposes itself on individuals with both banal and lofty explanations and expectations of femininity and masculinity that cause complications. Hence, masculinity and femininity are not products of biology, or not directly, as neo-Darwinians (evolutionary psychologists, sociobiologists and others) would argue, and hence do not proceed from DNA to the individual to society but from a cultural set of symbols (a superimposed structure) asserting its pressure downward on societies and individuals.

At the banal end of these theoretical positions, it is possible to see everyday translations of culture into gender-specific images and ideal types. The "ideal" woman and the "ideal" man, so Frigga Haug (1987) has argued, have been sexualized throughout the twentieth century. She maintained that the female body in particular has been sexualized and she examined the way various assumptions about sexuality and other qualities have been inscribed on the female body. One of her projects was to engage women in memory work, concentrating on specific parts of the subjects' body that had caused them some difficulty. Indeed, there hardly seemed an element of the female body that had not been subjected to the utmost scrutiny. One item concerned hair: bodily hair, facial hair, hair under the arms, on arms and legs, hairstyle, and length of hair. In all respects, and independent of the age of the participants, hair had caused problems somewhere on women's bodies and at some time during their lives. This was so either because hair was in places where it should not be or because there was not enough of it on the head, or it was not curly enough, or not the right color. While her subjects were all European, hair in females is an important marker. Hair tied into a bun, shaved off completely, or hidden has been a requirement of the married woman in many cultures, while hair worn open and loose has been regarded as an act of rebellion or sexual promiscuity. Thus writes one of the few female poets in early nineteenth-century Germany:

> Were I a hunter riding free
> Something approaching a soldier,
> Were I a man, if this could be,
> The heavens would lend me succour.
> Yet here I sit, delicate, fine,
> A dutiful child acquiescing.
> My hair I loosen in secret, alone,
> In the wind it flies streaming and flowing.
> (Annette von Droste-Hülshoff,
> cited in Haug, p. 104)

Legs, buttocks, posture, and breasts were sexualized and had to be a certain shape, size, and position, as also emerged in Haug's study.

We do not wish to expand these observations too far here other than to point to the extensive literature showing that women, and increasingly men, are tightly coerced into feelings of inadequacy about their bodies (Suleiman 1986). The images created and perpetuated through the media suggest a perfect femininity and a perfect masculinity that rarely occur, and so can be used against women and men (Wolf 1991). The idea of "perfection" tends to be in the direction of conservative gender roles and thus tends to perpetuate gender norms that some (or many) men and women may wish to shed.

The polarized presentations of particular concepts of male and female beauty mean to create a substantial distance between the real and the ideal. In doing so they suggest that a cosmetic product or any number of other products may be able to narrow some of the gap. Presentations of beauty are not just conservative. For the majority of the population, they are also unreachable because the stereotypes of the perfect male and female are incongruous with the sexual differences in physique typically encountered in humans.

To take Neumann's argument further: the *violation* begins with the body and through the body damages the personality. The body as a sexualized, idealized object and product, signaling success and health, has spawned many new problems, some of which are gender specific. Anorexia and bulimia, until recently, have been regarded mainly as problems of young women (Probyn 1988). Increasingly, the need to diet (and starve) has reached children not yet in their teenage years (Gordon 1998). Relentless cultural demands for perfection, performance, and unblemished appearance (from good skin, hairless arms and legs, fit and shining hair, no fat or wrinkles anywhere) have pushed up the stakes and raised expectations of appearance to new heights. Suicides and many other forms of self-injury have become far more common among young people. Self-endangering behavior is likely to be on the rise, and increasingly in men.

In the Western world, there now exists a vast and thriving industry to correct body parts that do not conform to specific standards of beauty and health. Surgery is often performed to make the physique of individuals conform to perceived ideals of how a male or a female ought to look. In most cases, women are the patients, having their breasts altered, their buttocks changed, fatty tissue removed, noses shortened or straightened, lips made fuller or thinner.

Apart from sexual attractiveness, appearance is meant to signal reproductive ability. Because of the latter emphasis, aging, particularly in women, appears to have been taken off the list of existing human conditions altogether. We seem to want to live in a permanent state of youthfulness (and that, too, has no biological basis). Ideas of dignity with age, the symbol of white hair as wisdom and equilibrium, are disappearing, presumably because it also signals the end of reproduction. One might add that, peculiarly, it is in an age of rampant overpopulation that the idea of reproductive ability has been raised to a new yardstick of desirability, both culturally and sexually. Fewer women of an age when it is natural to have gray hair now display it. Women, and increasingly men, color their hair to maintain a youthful appearance. Facelifts or partial repairs to faces have become more and more widespread, including the removal of extra folds in drooping upper eyelids. The purpose of these correctives is usually one of two things—either to look younger or to appear more sexually desirable. In both cases, a corrective usually also offers a psychological uplift for the individual. But the sense of deficiency or imperfection was planted in the individual's mind by a relentless media machinery designed to sell products. There is, and always has been, another dimension to aging and that is fear of dying. In the period of the Rococo, so great was this fear of visible aging, that wigs were worn by all ages and these were white-haired, the opposite solution to today's trend.

In linking physical appearance to fitness, in the Darwinian sense, another major travesty has occurred and that is the reduction of sexuality, of erotic, aesthetic, and complex psychological impulses, to reproduction. This signals a return to a simplistic kind of thinking based on biology. Many psychological schools of thought, including the Freudian, see sexual desire as a nonbiological constitutive phenomenon that evolves on the basis of the anaclitic principle related to needs and is quite separate from object fixation (Hoogland 1999, Laplanche and Pontalis 1988). It is important to bear in mind that sexual development may not coincide with, or be part of, a mere preparation for an appropriate gender role that, in turn, would be required for mating success. One of the greatest distortions is the idea that all physical organisms drive to reproduce (Dawkins 1976). In the natural world, only a fraction of all individuals in any species reproduce, the size of that fraction varying with the species. Yet some current biological theories imply that reproduction is (or ought to be) the underlying aim of all humans, a proposal which, in turn, demands attention to categories of feminine and masculine attractiveness. Instead, we ought to say, with Jessica Benjamin, that if we think of sex and gender as oriented to the pull of opposing poles, then these poles are not masculinity and femininity. "Rather, gender dimorphism itself represents only one pole, the other pole being the polymorphism of all individuals" (Benjamin 1995, p. 141).

We have not fully understood why psychosocial development suddenly poses so many new risks and problems in the twenty-first century, and it is of little help to see the flourishing of new scientific theories that simply dismiss these phenomena on the basis of genes. We do believe, however, that sexual reductionism is not only impoverishing culture in general, it is also creating new taboos and new repressions, some of which seem to weigh particularly heavily on young people. While we cannot examine this vast literature here, the fact that many girls develop a strong dislike for their bodies and a compulsion to stay slim to the point of physical harm and even death (Bell 1985, Bordo 1988, Gordon 1998) suggests that we have achieved little in liberating women from some perceived iron cask entrapping their bodies.

Certain traumas and conditions are more typical of males than females. Gender identity disorders, alcohol-related disorders, antisocial personality disorders, pyromania, and pathological gambling are found more often among males (American Psychiatric Association 1994).

Depression; anxiety; eating, somatic, and dissociative disorders are more prevalent among females (Brown and Finkelhor 1986). At no stage does the organism develop without substantial value placed on the genetic and apparent sex assignment. It is simply wrong to surmise, as Campbell (2002) has done, that these attitudes do not impinge on the way our gender identity develops. They do so from a very early age.

The pathology of this new sense of bodily and personal inadequacy is part of the dynamic of sexual repression that Foucault (1978, 1985) has described in detail. It is telling that, in his theoretical conclusion, "repression" does not work by silencing and tabooing a subject. In fact, it proceeds not by concealment and silencing but through exhibition, observation, and classification so that the very things to be repressed become depersonalized and reconstituted as an object, abstracted from the individual, incorporated into rules and schematized systems that appear as structures external to the individual, but yet are reimplanted within the individual. Hence, according to Foucault, the avenues of encroachment upon the body and on pleasures are multifarious.

8

Attractions

In this chapter we ask, "What sex differences, if any, might there be in feelings (emotions) and in the expression of sexuality?" This might seem to be a simpler question to answer than the controversial one about sex differences in cognition, covered in the last chapter. But this is not so. Sexuality is, in fact, the most contested research area of all. A whole host of theories link all of our patterns of behavior to the affective domain (to emotions) and therein to reproduction and biology. On the other hand, some schools of thought on these matters have been appalled at the very notion of biological determinism and view much of our affective domain as psychological and thus, by implication, as a cultural construct. The concepts of femininity and masculinity enter into the latter but as a cultural entity (as in Freud), not as a biological, or not necessarily even as an "essentialist" category.

ASSUMPTIONS AND HISTORY

Reproduction

One important domain in the field of sexuality is, of course, behavior related to reproduction. We can make the assumption, at our own risk, that sexual intercourse for reproduction is the most natural and fundamentally biological force. At the level of the individual, this assumption is not necessarily true. In the natural world, not all individuals in a given species reproduce and, at any one time, only a minority even has a partner. Only a few mammalian species form lasting bonds between individuals. Our closest relatives, chimpanzees and gorillas, live in large multimale, multifemale groups and females get "served" when they are in estrus. For many species, it is only during this time that male–female partnerships form, and in many cases, even this is very limited, with the sexual act the only direct evidence of a bond between them. Orangutan males disappear after a short period of consorting, leaving the entire duration of the pregnancy and the lengthy period of raising the offspring to the females. In fact, maternal orangutans may not engage in another sexual encounter for seven or more years (see Kaplan and Rogers 2000). Gorillas and, especially, pygmy chimpanzees (bonobos) often use sexual interactions as forms of appeasement (de Waal 1982, de Waal and Lanting 1997). Although long-term individual attachments defined by status in a group are known to occur, male and female pygmy chimpanzees are very promiscuous and much of their sexual activity is not for reproductive purposes. Apart from humans, we find the largest number of species that form pair bonds among birds, some of them bonding for life. Even so, only a relatively small percentage of birds within a species reproduce. Reliable research evidence now shows that only a quarter of most avian species breeds in any given year and this quarter is usually capable of replenishing the entire bird population (Kaplan and Rogers 2001). Hence, population numbers are held in equilibrium. Of course, for various reasons, the sizes of some populations are diminishing and others are increasing, but it is only the human population that has been growing continuously at an exponential rate.

Since breeding is not common to all animals, it is not a natural "law" that reproduction is the goal of every individual animal. Furthermore, one should not assume that those that breed successfully are always the healthiest physical specimens because there are many factors involved, not all of them related to skill, strength, health, and attractiveness. Bad luck, erratic climatic conditions, unexpected competition, death of a partner, or predator activity can rapidly change a successful breeder into an unsuccessful one. Who survives a storm is not always a matter of some genetic rule; it can be the luck of the draw. There are many factors that are incidental, surprising, and even random in life. Indeed, very extensive studies on birds have shown that reproductive success over the life cycle of many species studied could not be attributed to specific qualities of the mating pair, or of either of the individual partners (discussed in detail in Kaplan and Rogers 2001).

These studies are important insofar as emphasis on the "selfish" gene creates the impression that all life (meaning every individual) is always striving to reproduce and is driven to do so by the genes. We have absolutely no evidence for this. We cannot tell, for instance, whether a bird chooses to remain in a "bachelor" state or is forced to remain so by competition. We cannot tell whether bachelor groups are "happy" in themselves or whether they are just waiting for their chance to reproduce. It is a human assumption that the latter state is a state of waiting. Many fights are conducted among animals for the protection of a food source, so there does seem a strong desire to stay alive and, judging by the individual struggle of any animal threatened with death or loss of food, this may be universal. It seems that staying alive and reproducing are two very different modes of action.

When we see male or female rivals fight (competitions occur among both sexes), we are witnessing a specific group of individuals. There is no reason to assume that all seek to compete. Of the five million species of extant animals (invertebrate and vertebrate) we know something about only one percent of them, and of that one percent we have detailed research data on probably only ten percent. Therefore, it seems to us to be somewhat premature to assume laws of reproduction as driving the "system" of life on the basis of genes and their unbridled, unchallenged power to drive evolution. This is anthropomorphism, and specific, one might add, to capitalism, to project a sense of competition onto all living things. It is interesting that early animal studies undertaken under socialist regimes (or with a bent toward communitarianism) have emphasized collaboration and maintenance of harmony of a social group.

Reproduction in these studies often plays a very minor role. The current undue emphasis on reproduction (output) and competition (will to win) are so distinctly hallmarks of late capitalism that one ought to proceed with caution before making global announcements about the way nature supposedly works at the level of the gene.

In many ways, the new wave of genetic explanations of human behavior and human sexual behavior may be a new form of cultural repression. If we are driven by our genes, then it is the genetic action that requires explaining, while emotions, desires, and complex human interactions can be largely disregarded. We can describe stimulus–response pairs signaling reproductive intent and how mating takes place in, for example, stickleback fish. It is far more complicated when we get to birds and mammals. Breeding programs for apes and many other species have often failed in captivity because emotions and preferences were not taken into account. Male and female apes may spend years together in the same enclosure without a single mating attempt being recorded (Cantor 1993). The complexities of likes and dislikes and of fine discriminations of age and kin that may well be unrelated to health suggest that emotions may be involved and that they may tilt the animal into a nonreproductive mode.

Turning now to the human species, the frequency of marriage and the percentage of the population being married have gone up steeply over the twentieth century in most Western industrialized countries. Around 1900, as many as 25 percent of the population at the age of 44 years were not married and had never been married. By the end of the century, this percentage had dropped to a mere 11 percent, a remarkable increase in marriage rate while the rate of children born outside of wedlock has also increased. The human species has gone into overdrive to reproduce; research and medical energies have especially gone into reproduction at the historical moment when the human species has dangerously overpopulated the planet. If a species has ever been selfish it is clearly the human species, but one doubts whether this is a natural law.

Love? Girl Meets Boy

We might also query another set of assumptions that arise within the simple human tale of love and marriage: girl meets boy, they fall in love, get married and have children, and live happily ever after. The tale is specific to a particular time in history and belongs to specific societies. Yet, no branch of human endeavor, including that of science, remains untouched by its implications. Let us explore the assumptions underlying this beguilingly simple story, played out time and time again in novels and movies in Western societies.

First of all, the idea that a boy needs to meet a girl in order to fall in love is a heterosocial assumption not borne out by a substantial group of humans, as is now well known. Mammals, including humans, and birds have homosexual relationships and form long-lasting same-sex bonds; in fact, same-sex bonds are a good deal more widespread in animals living in the wild than is often reported (Bagemihl 1999).

The second assumption is that falling in love precedes the decision to get married. This has not been the standard practice in many parts of the world and over most of human history. The ideas of love and of happiness *in marriage* as prime or sole reasons for a union are eighteenth-century European inventions (Kamenka and Erh-Soon 1978); most societies found these preconditions for marriage decidedly odd. The same is true for some societies today. Many had, and some still have, arranged marriages, and the newlywed couple may not even be allowed to view each other closely or get acquainted with each other before the wedding day. Other marriages were effected to keep property, social influence, or power together. Terms of duty, obligation, and manipulation were/are more appropriate than love in these cases of marriage.

The third assumption is that marriage regulates sexual needs and thereby socially contains potentially irrational emotions and practices. The tale assumes that there are no other contenders for either individual and that acts of jealousy during courtship or marriage by a third party would not rip the happy scene apart. It also contains an assumption that "falling in love" progresses naturally to marriage and that this union is condoned by the wider community. But this is often far from the case in situations of self-selection of partners. Added to this, the tale assumes that the children resulting from the marriage will be genetically the children of the social parents, whereas genetic matching shows that they are often fathered by a lover (Coleman and Swenson 1994).

Yet another assumption is that the male and female know automatically what to do to conceive children. However, many childless couples attend family planning clinics even to this day wondering why they have remained childless, even though their union has never been consummated. It seems that there is no logical progression through meeting, bonding, and then reproducing. Reproduction may be seen as a by-product of the union or as the chief purpose of the marriage. In some cases, it was of political importance. Until the late Middle Ages and early Renaissance decades, lawyers witnessed the consummation of marriage on wedding nights of kings and princes. No one felt particularly strongly about a need for privacy in this regard.

The "boy meets girl" tale also implies that the girl and boy are not related. Some countries allow marriage of first cousins but marriages or sexual unions between brother and sister fall under the incest taboo and are considered a crime. Ancient Egyptians and Greeks, particularly within royal families, regularly practiced marriage between brother and sister but it is absolutely illegal in modern cultures, presumably because of the West's Judaic tradition in which any kin relationship even thrice or four times removed has been a punishable crime from the time Moses delivered the laws (Andreae 1998).

Finally, the choice of a sexual partner is, in fact, more complex than at first apparent. It may involve taking into account personality, conversation style, the context of the meeting, and many other factors in addition to physical appearance that will give us information, almost instantaneously, about age, status, life style, character, and even, at times, about the partner's political views. Hence, sexual and other forms of attraction and "falling in love" involve higher cognitive processes and decisions that may be evoked even at the first encounter. Sexual attraction involves choice, a response to a set of stimuli, a possible mix of markers that are pleasant from memory or experience and a range of nonverbal and verbal cues that might combine into the ill-defined and not well understood phenomenon of sexual attraction.

Sexual Expression and Gender Identity

There have been assumptions that sexual desire is a universal category and can therefore be discussed for all humans alike. The problem is that sexual desire does not exist outside the psychology of the human being and can therefore not be divorced from the entirety of the individual's affect and culture, which are certainly not divorced from morals and traditions. Attitudes to sexuality are shaped by laws; religious and moral values in particular may guide our concepts about our bodies and their sexual self-expression, both negatively and positively. Traditions change; it is unlikely that current sensitivities would allow lawyers as witnesses during the wedding night. Ernst Bloch (1985) said in the 1920s that human psychology is a historical category because it comes under a regime of contemporary cultural norms and sanctions. This may apply to sexual more than to any other category of behavior. Even talking about sexuality, particularly by females, has been taboo for most periods of Judeo-Christian history, at least until the late nineteenth century. In Chapter 6 we have demonstrated that, when sexuality was finally discussed, women in one generation apparently had "a sexuality" while not in the next. Women's sexuality fell in and out of favor according to social mores and goals, which must have been quite dizzying for those who were alive throughout all three shifts of perceptions of women's sexual desires (e.g., 1900s–1930s). The environment is thus not only a context of the immediate experience of the world (i.e., exposure to the immediate physical and social world) but a broader historical and cultural experience.

Another assumption holds that sexual identity, sexual orientation, and sexual partner choice fit together like hand in glove. Recent research on human sexual variability, together with research on biological and social variations, has revealed the falseness of the idea of a perfect fit between the individual's sexual identity, sexual orientation toward another person as a potential sexual and/or life partner, and direction of sexual expression (Evans 1993). Such complexity is frequently ignored when sexual behavior is studied in the clinic or in the laboratory.

SEX HORMONES AND SEXUAL BEHAVIOR

Much of the information about the role of sex hormones in the sexual behavior of humans has been extrapolated directly from experiments on rats and a limited number of other species. Those working on human sexual behavior often forget this fact, and so generate problems of understanding and of considerable magnitude in the social arena.

We know from experiments using rats that sex hormones circulating in the bloodstream affect sexual behavior by acting on receptors in the preoptic region of the hypothalamus (summarized in Breedlove 1992). These hormones may have a similar effect on the equivalent region in the human brain, but higher centers in the cortex of the brain also affect how responsive the nerve cells in the preoptic region will be to the hormones. As far as we know, these higher influences are far more important in humans than in rats. For example, testosterone has a more direct effect on sexual behavior in the rat than it does in primates, including humans, as shown some time ago by contrasting effects of castration. If an adult male rat is castrated, his sexual activity declines. It can be restored to normal by injecting testosterone (Larsson 1978) or by implanting a very small amount of testosterone into the preoptic region of the brain (Blaustein and Oster 1989). If an adult male rhesus monkey is castrated, he shows little or no decline in sexual activity, even though his testosterone levels become extremely low (Goy and Goldfoot 1974; see also Ward 1992). As no direct relationship between the level of testosterone circulating in the blood and sexual behavior occurs in the rhesus monkey, it seems that in more complex species, sexual behavior is emancipated from direct control by sex hormones, and higher brain centers are more important than the preoptic region. Although there are few rigorous studies examining sex hormones and sexual behavior in humans, the situation in humans appears to be the same as in the rhesus monkey: sexual libido of human males is relatively independent of the amount of testosterone circulating in the blood. In humans, castration does not lead to a consistent lowering of sexual desire and activity (Carter 1992), eunuchs may have been used to guard the harem but many of them engaged actively in sexual behavior. Similarly, men castrated for medical reasons in adulthood maintain normal levels of sexual activity (Swyer 1968).

The situation is much the same in females. Estrogen and progesterone are effective in stimulating sexual activity in rats; this sexual behavior coincides with estrus, when ovulation occurs. In primates, as we know particularly from research on rhesus monkeys and marmosets, but more recently from observations of pygmy chimpanzees in the wild, sexual interest and activity are not simply dependent on the estrus (or menstrual) cycle and therefore not entirely dependent on the level of hormones (Fabre-Nys 1998). When a chosen male is present with the female, mating may take place throughout the entire estrus cycle, although it may be slightly higher at around the time of ovulation. Castration has little to no effect on the level of sexual behavior in female rhesus monkeys, although a slight decline may occur in the long term (Wallen 1990). The presence of a preferred partner is paramount. Hence, the level of performance of sexual behavior depends on forming partnerships rather than on the levels of sex hormones in the bloodstream. The same is true in human females; social relationships have a stronger influence on sexual behavior than do the levels of sex hormones.

Testosterone or other male sex hormones, collectively called androgens, modulate the level of sexual interest in women. Androgens in women come from two glands, the ovaries and the adrenals. As we have said before, secretion of testosterone from these sources in women can even exceed the secretion of testosterone from the testes in males. Women have been treated with androgens for a wide range of medical conditions; even low doses have been reported to increase sexual interest. The decline in sexual behavior often following an operation involving removal of the adrenal glands can be reversed by treating the patient with testosterone but not with estrogen (Carter 1992). Indeed, in terms of sexual behavior, women are more sensitive to the effects of androgens than are men (Sherwin 1988).

All of these complexities make it impossible to tie any individual's sexual behavior to any particular hormonal condition, although some researchers try to do so. Apart from the important role of experience and the perception of stimuli from the environment, it is not simply a matter of knowing which hormones are circulating in the bloodstream. The number of receptors for each hormone in the various parts of the brain is an important link in the chain of events, since it is the receptors that recognize hormone molecules and remove them from the bloodstream so that they can work in the brain itself. If there are only a few receptors for a particular hormone, even high levels of that hormone in the blood will have relatively little effect in the brain.

The number of receptors varies from one part of the brain to another and also from one individual to another. Also, the sensitivity of the receptors can vary. No amount of hormone in the blood can make up for having insensitive receptors. Take the case of insensitivity of the receptors for testosterone, a condition referred to as androgen insensitivity. A genetically male person with androgen insensitivity does not develop facial hair or any of the other physical attributes of maleness even though he produces adequate amounts of testosterone. In addition, there are coactivators inside the nuclei of cells that modulate the effects of the steroid hormones on gene transcription (Auger et al. 2000).

Added to these complications, the hormone molecules in the blood can be either free, meaning that they are separate molecules not bound to any other molecule, or bound to protein molecules. Only free hormone molecules are available to bind to the receptors inside the cells and so only free hormone molecules can be effective in the brain. There is a balance between free and bound hormone in the blood—more bound molecules become free as the brain tissue removes free molecules from the blood—but both the amounts of bound and free hormone vary from one individual to another and from one stage of life to another.

Therefore, it can be seen that the physiology of the sex hormones has many and various facets and that no simplistic attempt to associate the blood levels of the sex hormones with any aspect of sexual activity is likely to be fruitful. Claims of showing such associations have been made in the past but they have either been distortions of the experimental results or based on flawed methodology. We can say that the physiology of human sexual behavior is far more complicated than scientists ever expected and that our current knowledge does not come close to understanding the "girl meets boy" scenario.

"MATING" STRATEGIES

The "girl meets boy" story in the sociobiological version is focused on mating. It presumes a "normal" male or female, attracted to a member (or members) of the "opposite" sex, who will eventually reproduce. In some versions, taking a spouse is implied; in other versions, particularly those of evolutionary psychologists, human males would benefit from polygamy. Sexual attraction to the opposite sex is supposed to work "naturally," and reproduction, a biological event, is purported to be a dominating drive that will assert itself for the entire species. The Darwinian idea of the survival of the fittest had new impetus with Dawkins's (1976) model of the "selfish gene," driven to replicate itself into the next generation at all costs. We have already said that it seems extremely unlikely that this is the case. Not all individuals show signs of striving toward reproduction, and it seems that some genes might not replicate themselves into the next generation. However, if this theory is believed prima facie, all behavior is the result of this requirement of the genes to replicate. In this context, physical features of males and females, and not just those of the reproductive organs, become important because they are said to be linked to reproductive practices, and evolutionary adaptations are made to fine tune and enhance gene replication. Humans are seen to also make mate choices and there are a number of rituals and assumptions by which these are allegedly guided that can be equated to the practices of any other vertebrate species.

Topics that concern sociobiologists include mating (particularly in terms of competition for mates), care of young, location of young, and practices to protect offspring. The theoretical basis for investigating human mating is taken from studies and insights into animal behavior. Hence we need to elaborate a little on the theoretical models of animal mating systems first in order to follow the train of thought taken in sociobiology and in evolutionary psychology. (For more details on the evolutionary psychologists' view, see Palmer and Palmer 2002.)

Sociobiology and Mating

Competition between males for a preferred sexual partner is said to lead to the evolution of extraneous physical traits (known as ornaments) and behavioral features of the male that attract the female (e.g., the beautiful tail of a peacock or the elaborate calls of the lyrebird). The more such competition there is between males, the greater the physical difference between males and females of the species (for details see Bradbury and Vehrencamp 1998). Group size, availability of resources, absence or presence of bonding and bonding focus, monogamy or fleeting consortships are all said to influence the way in which certain features of males and females have evolved. The resulting differences between the sexes, referred to as sexual dimorphism, are said to have signal function and may determine mate choice.

Among birds, there are examples of nearly absent and very strong sexual dimorphisms, according to the social context of mating. Monogamous, pair-bonding couples often do not have strong dimorphic traits, particularly when care of the young is shared. Albatrosses and ravens, for instance, mate for life, and in both species sexual dimorphism is minimal. In such species, the identifiers of male versus female may be as minute as the color of the eyes (as in galahs), the color of the cere (as in budgerigars), or the color of the tongue (as in ravens—it is black in dominant males).

If males have no part in raising their offspring and will mate with many females, their sexually dimorphic attributes are usually expressed in dramatic plumage color or plumage effects, specialized song or dance, nest building skills, or in a combination of some or all of these. These attributes are said to have developed because the male has to attract a female. He may need to impress her because once he has mated with her his input into incubating the eggs, and feeding and protecting the offspring may be minimal or even absent. Strangely, birds with this kind of parenting have generally attracted far more interest in research than those in which the male has more parenting input.

In a small but significant group of bird species, the male, it seems, will offer no more than his special attributes before mating and literally nothing thereafter (Kaplan and Rogers 2001). Offering as little as he does, the females are said do the choosing, and hence some adult males compete fiercely, trying to outperform each other. Some males have almost too many offers while others have none. Over time, the preferred trait or traits that may be instrumental in having high and consistent reproductive success may become so exaggerated that this results in a "handicap principle" (Zahavi and Zahavi 1997). The train of the peacock is a typical example. As beautiful as it is to our eyes, and as attractive to peahens as experiments show that it is (Petrie et al. 1991), it severely hampers the bird's ability to move around and to escape from predators. In birds of prey, there are strong size differences; here it is the female who is substantially larger than the male, a trait that may be an adaptation to the needs of caring for offspring. It may be that her formidable size will prevent him from consuming his offspring and/or it may be meant to ensure that he is easily enticed into giving up his prey. Raptors always rear their young together as a pair. This may be necessary because of the greater difficulty of securing food by hunting as opposed to food procurement by insect, fruit, and seed eaters. There are also some mammals in which the female is larger (Ralls 1976), but no research data is available that can explain their larger size, as if to suggest that, by not fitting the sociobiological explanation, these cases (and they are not insubstantial) are of no interest.

Female mate choice in birds is also based on another evolutionary development of particular relevance to male mate competition. This is the female's ability to abort and the need for her to cooperate in mating, so that the cloacae of both the male and female birds can meet. One notes that sociobiologists almost always disregard this aspect of mating behavior. It is the fact that most male birds do not have a penis and that the female can abort an unwanted egg very swiftly. If a male has forced himself onto a female and has been rejected by her, he will usually not succeed in having offspring. There are only few examples of forced intercourse in birds. These are almost exclusively in waterfowl, where a penis has been retained (McKinney et al. 1983).

In mammals, there is rich evidence of sexual dimorphism, although in some species it is not strong, especially in some group-living mammals, such as dogs (Rogers and Kaplan 2003). When it exists, it may entail differences in body size between male and female, in weight, in voice, in fur thickness, in color, and in the absence or presence of additional features (such as horns, manes, cheek pads, or beards in males). Male-male competition is said to increase the size of the male compared to the female. There is certainly plenty of evidence among various species of seal, especially when the males hold harems and spend much of their time fighting off other contenders. After a few years of this highly stressful existence, the male is usually so exhausted that another bull easily wins the fight and will take over the harem. For all his size, his dominance and possession of a harem is very limited and, as in other species, many never get to mate at all, instead spending their lives in large bachelor colonies.

Sex differences in weight, substantial among some primates, has been related to sexual selection (Gaulin and Sailer 1984). In chimpanzee and gorilla societies, consisting of a dominant male, and some younger males, and a number of females, the alpha males are substantially larger than the females. The strong hierarchy between chimpanzee males (not necessary females) results in significant size differences between the sexes to ensure that the dominant adult male has exclusive rights to the females of his group (de Waal 1982). In orangutans, males are twice the size and weight of females and are strongly dimorphic, with other secondary sexual attributes in males only, such as the cheek pads and the air sac for specific vocalizations (Rodman and Mitani 1987). Males form only brief consortships with females and then leave pregnant females and have no further input (Nadler 1995). Subadult orangutans may only gain some occasional sneak mating opportunities when the adult male is not looking, but they would not dare compete with him in his presence. The imposing side flanges (cheek pouches) that mature orangutan males develop may well function as another deterrent, since immature subadult males develop these flanges only when there is no mature male in their vicinity (Kaplan and Rogers 2000).

Generally, it is assumed that the more egalitarian a group structure is, the more pronounced and exaggerated are the primary sexual traits (the genitals). This trend tends to hold for mammals, but not for birds. Among the great apes, chimpanzee and orangutan males have small penises and small testes but pygmy chimpanzee males, living in a more egalitarian and matriarchal society, have the largest testes of all great apes (Thompson-Handler et al. 1984). Human males may have among the largest penises in relation to body size of any primate. It is therefore thought that these evolved because of a more egalitarian group structure, or possibly, the sociobiologists say, because ancestral humans were matriarchal, as in pygmy chimpanzees, our closest relatives.

Relevance to Human Mating

For some time, physical anthropologists have puzzled over the sexual dimorphisms among humans and have asked whether they have been caused by sexual selection based on certain mating principles. For instance, according to the literature, human males are 4 to 10 percent larger in body size than females, regardless of climate, ethnicity, or diet; this difference deserves explanation (Hall 1982), be it mainly a genetic issue (Rogers and Mukherjee 1992) or a result of social experience.

Studies of animal mating systems and the assumed relationship between specific selection pressures and the evolution of certain sexual dimorphisms, as we have discussed, led sociobiologists to ask about associations between the magnitude of sexual dimorphism and the form of mating system in humans (Miller 1998, Trivers 1972; see also Palmer and Palmer 2002). On the strength of this, Gaulin and Boster (1992) examined stature differences in 155 societies and established statistically that polygyny in nonstratified societies is associated with a trend toward higher sexual dimorphism in stature. However, the difference in stature was, in fact, not substantial enough to lend strong support to the sexual selection theory. This could be so because, as they argue, human marriage systems are "relatively transient phenomena and may rarely persist long enough to measurably alter the magnitude of sexual selection" (pp. 473–474). This finding is not surprising and emphasizes the fact that human practice is very variable. Despite the existence of rules and sanctions, these practices are often intercepted and the rules broken, or the rules and practices may change slightly from generation to generation without affecting the general body of laws and rules of a society.

In this focus of research, equating animal and human mating systems and trying to discover the same principles as they pertain to animals, sometimes somewhat bizarre hypotheses have been put forward to explain sexual dimorphism in humans. For instance, Morris (1967) and Gallup (1982) sought to explain why human females have "permanently enlarged breasts." Morris attributed enlarged breasts to the evolution of bipedalism (walking on two feet) and face-to-face mating. He argued that the enlarged breasts are a replica of the buttocks used as a signal for sexual attraction. Gallup has suggested that breast enlargement may serve as advertisement for ovulatory potential, and so compensate for the fact that ovulation is concealed in humans. From this position, he then developed a global explanation for sexual preferences in men. This hypothesis has, of course, not been tested, but such pseudoscientific reasoning at times grows into generalized claims with broad social implications. No attempt was even made to refer to the relevant primate literature. It may be pointed out that ovulation is also concealed in orangutans and they, too, practice face-to-face mating and have sexual intercourse in about the same manner and for about the same duration as humans. Pygmy chimpanzees often engage in face-to-faace mating, but they do not conceal ovulation. Like humans, both orangutans and pygmy chimpanzees copulate throughout the menstrual cycle; yet neither pygmy chimpanzees nor orangutan females have enlarged breasts (see Kaplan and Rogers 2000).

"BEAUTY" AND SEXUAL SUCCESS ACCORDING TO SOCIOBIOLOGY AND EVOLUTIONARY PSYCHOLOGY

Mating success has also been tested in conjunction with other markers, and here we get into the cultural domain of beauty. Again, the underlying assumption is that "mating success" in humans can be described in terms of genetic ideas about sexual selection. Hence sociobiological researchers have investigated the nature and function of physical attractiveness, as this was thought to contribute to mating success just as ornaments are considered to be involved in mating success in avian and nonhuman mammalian species. The fact that we have data on relatively few avian species and that we still may not know all the factors contributing to mate choice should be an obvious limitation to generalizations. Moreover, most of the mate choice paradigms refer only to species with relatively strong sexual dimorphisms. We know relatively little, and in most cases next to nothing, about mate choices of less sexually dimorphic species.

Yet, studies of human attractiveness have relied on that small pool of animal studies and have applied these principles to humans. Among humans, we have the advantage of not needing to rely solely on observation but can directly question the subjects. In such studies we know that a range of nonobservable criteria such as life attitudes and certain character traits (honesty, kindness, sense of humor) play a significant role in sexual choice, especially for women (Mirsky 1999).

Ranking someone else's attractiveness and counting one's chances to have such feelings reciprocated if an approach is made, also depends on the initiator's perception of his or her attractiveness. A study found that people who think they are attractive opt for a more attractive mate (more attractive in their own perception) than they would have chosen had they thought themselves less attractive. The study asked each group to rank another group according to a scale of attractiveness (Kowner 1994). They were then given bogus feedback that they were ranked top in attractiveness by the group and asked to make their final choice. Immediately, they aimed for a more highly attractive person (in their judgment) than they had chosen before.

One important finding by Kowner is the dialectic approach to a particular stimulus. The stimulus (another human as a potential partner) is not just one to whom one responds statically as in a simple stimulus–response pair. Indeed, the process is dynamic because the stimulus is not just evaluated with respect to its intrinsic properties (an attractive male or female) but it is first processed according to how the observer's intrinsic qualities rank with respect to the other. Only after that initial assessment is the judgment of attractiveness announced. Hence, the observer judges how likely it is that access to the other will be possible and thus makes the preferred choice (and the ranking of the other) according to self-assessment. Attractiveness is thus not only a question of the evaluation of another but also of oneself.

Yet evolutionary psychologists and biologists, as well as sociobiologists, want to make the link with genes rather than speak about any of these other accessible attributes. Wedekind and colleagues (Wedekind and Furi 1995, Wedekind et al. 1997) have described traits of health that are communicated by body odors and other cues, and that these signals affect mate choice. Body odors can be judged as unpleasant or pleasant, as attractive or repulsive and, although not consciously, may correspond to certain expressions of good or ill health. Wedekind and colleagues argued that the major histocompatibility complex (MHC) played a role in mate choice because it influences body odor and body odor preference. The odor preference of women for men is, apparently, dependent on women's hormonal status. The researchers reached the conclusion that genes associated with MHC, or linked genes, influence human mate choice.

But how important is olfaction, compared to the other senses, in mate choices made by humans? We already know from countless studies in primates (especially great apes, the evolutionarily closest relatives of *Homo sapiens*) that the importance of olfaction declined over primate evolutionary time in favor of vision and hearing. In the case of olfaction, there is no point, as Pawlowski (1999) has done, to place importance on the olfactory environment in rodents, since olfactory traits ceased to play the same important role in great apes. Pawlowski is also incorrect in stating that the female of the human species is alone among humanlike [sic] species in not having an estrus cycle that displays when she is ovulating. As we have mentioned, orangutan females also have concealed ovulation (Kaplan and Rogers 2000), suggesting a much earlier evolution of this characteristic, predating the hominids by eight to twelve million years. Although humans do relate to olfactory experiences and form olfactory memories (Schleidt 1992)—sometimes quite important ones (such as special perfumes)—the role of olfaction in non-human and human primates has not only substantially diminished over evolutionary time but, significantly, its role in sexual attraction may have diminished even more so. We simply do not sniff each other's genitals (nor do the great apes) in order to assess reproductive status and mating readiness. We admit, however, that some odors still matter and that others may have taken on this role to which we may often respond unconsciously.

It has been postulated by evolutionary psychologists that symmetry of physical features (of not only the face but many other parts of the body) leads to more mating success than asymmetry (summarized by Concar 1995). The left and right sides of the face and other parts of the body are asymmetrical to varying degrees. This is called fluctuating asymmetry and is found in all species. But Randy Thornhill and colleagues (1995) have claimed that symmetry is the main criterion of beauty and sexual success. Human faces that are more symmetrical in appearance tend to be seen as more attractive and more likely to attract a sexual partner (Gangestad et al. 1994, Thornhill and Gangestad 2000). The reason for this, it is claimed, lies in the fact that symmetry represents health and thus good reproductive potential. Asymmetries may result if a person has "bad genes," is carrying mutations, or has been attacked by a pathogen or a toxic substance during development. The same selection for symmetry is said to happen in a wide range of species, not only humans. In fact, Thornhill first formulated his theory when studying mating behavior of the Japanese scorpion fly. He found that males with the most symmetrical wings performed more matings than average.

The same idea was applied to other species, and some evidence in support of it has come to light, apparently even in humans. Symmetrical humans, according to Thornhill and colleagues (1995), are the beneficiaries of a greater number of orgasms than asymmetrical ones. The first evidence was obtained by selecting eighty-six heterosexual couples from among students studying an introductory psychology course and measuring the widths of the men's feet, ankles, wrists, and other parts of the body, as well as the sizes of their ears. The degree of left and right side asymmetry for each of these features was calculated and then these values were correlated with the frequency of sexual orgasm reported by their partners. The latter was determined by getting the subjects to answer a questionnaire about their sexual behavior, a notably unreliable source of information as known for some time. Women whose partners were the more symmetrical males reported that they had more orgasms during copulation than women married to less symmetrical males. This is not to say, however, that facial symmetry, equated in some of the literature with attractiveness, would necessarily be what is known as an "honest signal" of having "good genes" (Kalick et al. 1998, Shackelford and Larsen 1999).

The researchers interpret their result as support for the idea of sexual selection for symmetry in humans, as they had done previously with flies and other species. That is, they believe that they have support for their theory that symmetry and reproductive success is a basic principle of biology and that it has something to do with selecting the best genes for the next generation. We have to say most emphatically that there is no known mechanism that links human genetic fitness to symmetry or fluctuating asymmetry and that the connection is spurious, unless we include a whole host of cultural considerations that may explain the replies about sexual orgasm and the findings. The connection between female orgasm and increased likelihood of reproductive success has been made in a number of publications (Baker 1996, Baker and Bellis 1989, 1993). Physiological changes with orgasm are known to aid sperm retention and may thus assist conception. However, the work by Baker and Bellis was largely related to the role of the lover and thus at least hinted at love, an aspect of mating (in humans) that is profoundly missing in sociobiology.

Further criticisms can be leveled against the work of Thornhill and colleagues. For instance, even if their findings from scorpion flies or birds could conceivably apply to humans and were to be thought applicable generally in different ethnic groups, different socioeconomic classes, and so on, there are various ways of interpreting the data. If symmetry of the male partner really correlates with frequency of female orgasm, this does not tell us anything about the factors that influenced the making of this relationship. The causes might be various and complex.

Since the results of their study depend entirely on self-reported answers to questionnaires, and certain biases and distortions are notoriously common in such questionnaires, we might speculate as follows: Let us assume that people with greater symmetry have consistently had flattering responses from other people about their appearance and, in some cases, might have had these comments made to them or in their presence since they were children. They may have learned early that their appearance matters. Let us further assume that such comments about appearance were rarely matched by specific positive references to any of their character traits. These people therefore might have grown up believing that appearance is all that matters. Orgasms can feature highly in the world of appearances and pretence. There are social environments in which orgasm signifies apparent sexual success, as we have seen in Chapter 6. One generation can talk freely about orgasm and this may even become "cool," while another generation might have tabooed the very notion of it. If appearances count and orgasm is socially important, it can be used as a form of bragging (we are happy, we are virile, etc.). In this scenario, we would expect a much higher incidence of reported orgasms from more attractive people than from those who are perceived by others and by themselves as less attractive. Conversely, subjects with less symmetry and, presumably, a lower sense of their own attractiveness, may not have a need to make appearance count and they may report honestly or even underreport. This suggested interpretation would be in line with the results found in the study ranking attractiveness, discussed earlier in this chapter.

One could offer many theories, but the fact remains that the basis of many of these speculations is very shaky indeed. Research along these lines could be seen as attempts by sociobiologists to get into the field of sociology or social psychology with no methodology used to account for complex social and psychological forces impinging on the behavior under consideration.

To return to the Kowner study mentioned previously: There is another very important point arising from this experiment, and it is one that evolutionary psychologists tend to forget, whereas marketing departments in business enterprises would forget it only at their peril. Humans tend to construct ideal worlds for themselves and may, in wishful thinking, choose objects and people fitting those constructions. However, in reality, their choices may be quite different. For instance, a car company wanted to design a real market winner and spent considerable funds asking people's opinion on what kind of car they really wanted. On the basis of the results they built a car that turned out to be the company's most disastrous market failure. People may say they want a Jaguar but will actually come home with a Ford, even if they could have had either car at about the same price. The Jaguar fits the image that they might construct for their ideal world, but they would feel embarrassed to parade these dreams openly for everyone to see. Choice of partner according to criteria of attractiveness may suffer the same fate. People may have movie stars and rock stars as their dream image. They may place a centerfold of the perfect (stereotyped), utterly symmetrical, male or female beauty on the wall or in their locker, but in some cases might feel exposed and probably even uncomfortable, albeit titillated, if that person stepped off the centerfold and escorted them home to their family.

Genetic explanations for sexual attraction are part of a broader wave of opinion denying the importance of social and psychological forces in the development of human behavior. Links have also been made between facial attractiveness as a predictor of performance attributes and career (Mueller and Mazur 1996), but the latter has been seen as a function of a conscious or subconscious organizational image-making process and may have very little to do with sexual attraction.

There are many explanations for physical attractiveness in the social realm. After conducting extensive experiments, Mulford and colleagues (1998) came to the conclusion that physical attractiveness cannot account for more than a small part of the variance of success. They found that attractiveness contributed to a belief in the attractive person's willingness to be collaborative, and that when these hopes were not justified, there was a tendency to stay away from any other attractive person, conforming to the view that "beauty is only skin-deep." Even among evolutionary psychologists, the relative importance of attractiveness in mate choice is not always clear. In 1999, a study of personal advertisements looking for a partner in daily papers found that "market value" of males increased in accord with higher income and lessened with perceived "risk of future pair-bond termination." Women's market value was dependent on fecundity. For both sexes, market value decreased as a function of age (Mirsky 1999) but there was little direct evidence of physical features as a driving force in the desirability of a partner. The difference, so Bereczkei and colleagues (1997) explained, lay in whether or not the intention was to look for a short-term mate (without reproduction) or for a long-term mate. In their study, also of personal advertisements, sexual attraction played a role largely when only short-term commitments were intended (Bereczkei et al. 1997).

If this is so, then one might actually arrive at the opposite predictions from those that evolutionary psychologists make. From the female's point of view, mate choice, as a mechanism for reproduction and passing on genes, should favor a long-term mate and, according to these studies, attractions of symmetry should recede in favor of other qualities. In fact, according to some studies, this may be the case (Rhodes et al. 2000).

So far, most psychologists have considered sexual attraction only as a matter to be extrapolated from sexual practice (i.e., the response end) and thus have accepted, and mistaken, social norms for biological reality. As Hoult (1983–1984) has pointed out, the biological model is inappropriate to an understanding of human sexuality, be it in terms of sexual actions per se, sexual orientation, or gender identities. Sexual attraction has been channeled by most societies into a consciousness limited by cultural and social practices, although there are, of course, cultures that have placed less emphasis, or different emphases, on individual sex characteristics (Ford and Beach 1951). To mix "mate choice" and sexual attraction, however, and presume them to be the same, is very flawed. There are sexual attractions that do not lead to the passing on of genes, and there are mate choices that are made on grounds other than facial symmetry!

CULTURAL EXPRESSIONS OF SEXUAL ATTRACTION AND AMBIVALENCES

Sexual attraction is also more complicated for another reason, ignored by evolutionary psychology. Maleness and femaleness are not unitary concepts. We have argued elsewhere (Kaplan and Rogers 1984) that sexual attraction is stronger to individuals who show a mix of male and female physical characteristics and masculine and feminine behavioral traits. All people, to varying degrees, have a mixture of male and female characteristics, both physical and behavioral. Recognition of this fact raises the question of what exactly we mean by male and female characteristics, and indeed whether the terms are at all useful or appropriate. A male may respond to a female not only because some of her physical and behavioral attributes are "feminine" but also because some of them are distinctly "masculine," and vice versa. Intermingling of male and female characteristics may occur within an individual's physical features. This is not to be confused with androgyny, commonly used to indicate a fusion of maleness and femaleness into a unified whole by way of minimizing primary and secondary sexual attributes of both sexes. The concept of androgyny certainly has its place, but, if accepted as the only reference to a mix, obscures the many attributes precipitating and facilitating attraction of one individual to another.

It has long been known by artists of Eastern and Western cultures that the male–female mix can be exploited very effectively in just about any art form; it is not something that has arisen recently. Indeed, the sexual mix runs through the entire cultural history of Europe and the Western world. According to Greek mythology, Bacchus (Dionysus) was the god of fertility and ecstasy and later also of wine. Interestingly, he was always considered the opponent and competitor of Apollo (his half-brother), the god of music and poetry. Bacchus was a lover of men but women loved him with such frenzy that they celebrated orgies in his honor. This aspect of the Bacchus myth, namely his extraordinary ability to sexually attract men and women alike, must have inspired Caravaggio's painting of Bacchus. Unlike his predecessors, who depicted him as a bearded man or, in later centuries, as a young man wearing an animal skin, Caravaggio (1560–1609) chose to endow him with male and female attributes and to let him rest on a pillow in a rather seductive pose.

There is much evidence in Western and non-Western cultures of a strong interest in plots and characters involving costume disguise across biological sex. The theater is the place in which conscious illusions are created and it often takes the liberty of saying and exposing what might otherwise not be permissible. The enjoyment of the male/female mix falls perhaps most readily into this latter category.

Breeches or trouser roles for women (i.e., cross-dressing as men) and transvestite roles for men began to become part of the repertoire in Western culture in the sixteenth and early seventeenth centuries. Wandor (1981) argued that transvestite theater

> has flourished at times of changing attitudes to women in the theatre and to sexuality in society—the Restoration, the Industrial Revolution, the suffrage agitation and now, in the second half of the twentieth century. At such times clearly there is a tension between the surface appearance and how men and women are supposed to "be" and the changing reality. [p. 19]

This explanation is an attractive one in the sense that it highlights how theater may respond to social changes by experimenting with the extremes of such change, as role reversals would be. However, examples of cross-dressing are not confined to the periods Wandor specifies and are not less numerous, or at least not less important, in other periods.

Perhaps one of the best known examples of an early breeches role is that of Viola, traveling as Cesario, in Shakespeare's *Twelfth Night*, written and performed in 1601, and based on an Italian comedy of 1531 (Halliday 1964). Olivia falls in love with Viola/Cesario while Viola herself has fallen in love with the duke Orsino. The audience is aware of Viola's disguise and can therefore relish the confusion in the relationships of Cesario and Orsino, and Cesario and Olivia: Olivia pining for the disguised Viola and the supposed Cesario blushing and making eyes at Orsino. The narrow confinement to sociocultural roles has in fact not disappeared in the play: Olivia is capable of loving Viola but can do so only because she believes her to be a young, attractive male. It is not clear whether Olivia loves the image of Viola in men's clothes or of Cesario with feminine features. Luckily, Sebastian, Viola's twin brother, can be found in time and can replace Viola. Viola, discovered in her true sexual identity, can now proclaim her love to Orsino. Olivia can love Sebastian precisely because he is so much like her. The plays by Shakespeare were performed to a wide cross-section of the London public, including the poorer people, the well-to-do merchants, and the aristocracy (Salinger 1970). This suggests that crossing of roles, and hence mixing of physical types, held a very general attraction.

If we include ballet, opera, and film in the performing arts, the evidence of cross-dressing and double play with gender becomes even more pronounced. Opera, with its combination of acting and singing, was able to add the auditory to the visual stimuli and thus take the disguise, gender confusion, and cross-dressing a step further. In its beginnings, it did so by the introduction of castrati, singing at a voice level of soprano and contralto. Moreover, castrati sang both male and female roles. The castrati were very important figures in Western opera, right from its beginnings in 1607 when they first appeared in Monteverdi's *Orfeo* until around 1800 when their use was prohibited for humanitarian reasons. In these two centuries they often achieved prominence and were increasingly afforded great power (often political) and wealth. The castrati are said to have been impressive figures on stage, "tall and broad-chested" (Heriot 1956, p. 27), and they were extremely popular in roles, even tragic ones, in which they portrayed female characters. They were in such demand in the eighteenth century that as many as 70 percent of all male singers were estimated to have been castrati, in the absence of any compelling social reasons preventing female singers from appearing and in the absence of any shortage of female singers (Heriot 1956). Goethe (1749–1832), one of Germany's most eloquent poets at the time, made special mention of these castrati:

> I reflected on the reason why these singers pleased me so greatly, and I think I have found it. In these representations, the concept of imitation and of art was invariably more strongly felt, and through their able performance a sort of conscious illusion was produced. Thus the double pleasure is given, in that these persons are not women, but only represent women. [cited in Heriot 1956, p. 26]

When Silberberg produced a film of Wagner's opera *Parsifal*, he was obviously playing with this tradition but reversing the sex, a feat that could never be carried out in live theater. Here Parsifal, once having reached adolescence, is played by a woman but continues to be sung in a baritone male voice. Interestingly, *Der Rosenkavalier* by Strauss (1911) uses sex-role disguise in much the same way that Shakespeare used it in his plays. The part of Octavian is played by a soprano and Octavian disguises himself as a woman in order to entrap and expose another male character.

In ballet, the role of Franz in Delibes' *Coppelia* was danced by a woman when it was first performed in 1870. Tchaikovsky's *Nutcracker Suite* contains a pantomime role of the Old Woman who lived in a Shoe, to be danced by a male (Warrack 1973). Apart from these examples and the comedies that continued to use cross-dressing, the melodrama made its appearance on European stages at the close of the eighteenth century, at once including cross-dressing in its repertoire. In the twentieth century, the medium of film continued to exploit, often in novel ways, the presentation of the mix. The film historian Homer Dickens, in his book *What a Drag* (1982), has examined more than 200 cinematic treatments of cross-dressing in films from the 1920s onward. Today, plays and films continue to be written and produced incorporating cross-dressing (Engel 1985). Of course, cross-dressing was often used as a comic device in the theater. In this context, it neither challenged nor undermined traditional mores but reaffirmed them even if, at times, it may have contained the seed of rebellion against conformity (Hyslop 1985).

Eastern cultures also have well-entrenched theatrical traditions and cultural practices of cross-dressing and of the representation of the mix. For instance, the main characters in the kabuki theater are the female onnagata played by highly trained male actors. Even in present-day Japan these actors have a cult following. The kabuki theater originally developed in Kyoto in the early seventeenth century with an all-female cast. The word *kabuki* means "to be unusual" or "out of the ordinary," with the connotation of sexual debauchery and, indeed, most actresses were also prostitutes. An injunction of 1629 would almost have ended the kabuki theater had it not been saved as an art form by some determined males who took over all the roles (Ernst 1956), the female roles being played by beautiful young boys. These boys were so revered and admired that samurai fell in love with them and, after a public brawl between two samurai, this form of kabuki theater was also banned. The reopening of the theater was granted on the condition that the onnagata cut their hair, wear wigs on stage, and indicate to the audience by subtle hand movements that they were in fact males. It is in this latter form that the kabuki theater has survived to the present day. For Japanese audiences, the onnagata are special favorites, spurred on noisily with calls of enthusiasm, appreciation, and support throughout the performance.

In Chinese opera we also find female impersonators. Their history is considered to be longer than that of the onnagata, but the reasons for their appearance are likely to be different. Female impersonators have appeared in Chinese opera performances since ancient times, very often because emperors intermittently forbade stage appearances of males and females together (Mackerras 1972). This was particularly so in the eighteenth century (Alley 1957). The practice of males playing the female roles has survived to the present day, although the restrictions placed upon the sex of actors are no longer applied. Many Western observers are unaware of this fact. There are operas such as *The Iron Bow* in which we find the convoluted situation of a male actor playing the female role of Chen Xiuying, who then takes on the disguise of a man and goes to battle, overwhelming a battalion of troops in order to avenge her husband!

A woman playing a male pretending to be a woman was exploited in the film *Victor/Victoria*, with Julie Andrews in this role. But inbuilt in the story is a conservative resolution of the complex plot in that, in the end, Victoria reemerges and chooses a feminine future with her macho male lover.

In present-day theater or film, cross-dressing itself has at times become the very subject matter of a plot, as for instance in the film *Tootsie* or in such plays as *Torch Song Trilogy* or *Eugenia Falleni*. The latter was based on the true story of an Italian woman migrating to Australia as a stowaway at the end of the nineteenth century, disguised as a man. She lived as a man and married twice, murdering her first wife after the discovery of her true sexual identity. Eugenia Falleni was sentenced to death, but her sentence was commuted to life imprisonment. Later she was released and died, after a few lonely years as a handyman, in Sydney in 1939.

The persistence of images purposely portraying the attraction of the male and female mix is unlikely to be an accident. Although cross-gender disguises tend to occur in serious plots in Far Eastern cultures, and as stock-in-trade of comedies, farces, and melodramas in Western cultures, cross-dressing undoubtedly has an appeal to the audience, precisely because of the double message of the portrayal of the protagonists. It would be difficult to conceive of its continuing popularity through the ages otherwise. The pleasure is not derived from a fusion of male and female attributes into one perfect form (as in androgyny) but from the assertion of both, side by side. The intention of cross-dressing is ambivalence, bisecting categories that are often well fixed and reified in everyday life.

The experimentation with gender, with cross-dressing and with role-reversals, left the stage in the late twentieth century and became a social issue, not for the sake of entertainment but for the sake of self-assertion (Glickman 1985) or personal identity (Lewins 1995), and not just for outsider or minority groups but for any man or woman. It was driven into reality partly because of surgical and technological advances. Individuals can surgically change their sex, not just as a dress-up crossing over, but as a real anatomical change. For Jan Morris, a male-to-female transsexual, for instance, sex is a physical state while gender represents "the inner consciousness-abstraction, not anatomy" (cited in Allen 1984). In a transvestite photographic exhibition, one photographer (Allen 1984) writes: "And in the surgically advanced era in which we live, anatomy itself becomes an illusion. But the greatest illusion is that our sex determines how we are supposed to conduct our lives" (p. 1).

Such moves away from the discrete polarization of male and female characteristics show that sexual attraction is indeed a much more complex phenomenon than the simple male/female dichotomy suggests. The artist stepped in at the point of the puzzling incongruity between social norms and actual behaviors and intuitively formulated these irrationalities long before science began to systematically investigate questions of maleness and femaleness or of sexual attraction.

Despite repressive theories, and despite the fact that many scientists have been tardy in recognizing the lack of a sexually absolute dichotomy, "maleness" and "femaleness" and "femininity" and "masculinity" are not opposite ends of a polarity. This polarity has been a human construct, not one caused by biology, but a conceptualization *within* biology. On the contrary, had the human species not adapted to the reality of some degree of the mix, sexual attraction could not occur in most cases. However, sexual attraction is clearly not an area of common sense, nor, for that matter, are the constructs for social existence. Throughout history and today, the transcendence of gender identification, the purposeful confusion of expected elements of the male and female types in the theater, in fine arts, and in dance has had a large following, often by the same people who would, in real life, totally refuse to admit their fascination, even titillation with such a mix.

A study conducted by Rhodes and colleagues (2000) has found that super-maleness and super-femaleness, even when perfectly symmetrical, are not the most preferred types and do not receive the highest ratings in attractiveness. Coming from a very different theoretical direction than ours, her results corroborate our argument here. Indeed, male faces that had been feminized (taking elements from female faces) were considered more attractive than super-male faces; this preference persisted across cultures.

In 1975, P. Gagnon declared that the issue of what, in behavioral terms, constitutes a man or a woman is an open question. Meanwhile, his point has been expanded to include the phenotype (i.e., physical features), while the issue of *sexual* attraction, as distinct from attractiveness, which can have an entirely asexual and purely aesthetic quality, has been left largely unprobed.

We have purposely juxtaposed scientific and cultural explanations of sexual attraction to indicate that we are dealing with hierarchies of knowledge and explanations and with sexual, psychological, social, and cultural phenomena that inform each other. After centuries of puzzling about sexual attraction and bonding, it begs the question whether love is entirely explicable just in terms of "reproductive strategies" and goal-oriented gene intentionality, let alone being reduced to mere trappings that spin off from the basic purpose of the replicator units.

9

Gay or "Queer" Gene

THE SEARCH FOR SPECIFIC GENES

Many medical researchers have now cast their eyes on genes as possible causes of certain diseases. Some of these studies have been relatively successful and, in the last decade or so, have provided a clearer picture of certain inheritable diseases, such as Huntington's disease, cystic fibrosis, and muscular dystrophy. Not surprisingly, there is widespread optimism and support in the media and in the population generally for genetic research, even though, so far, these particular insights have not led to possibilities of manipulating or significantly changing the course of these diseases once diagnosed.

There can be no objection to this medical attempt to identify genetic abnormalities in association with debilitating physical conditions, apart from a consideration of cost at the expense of funding research on diseases that afflict larger numbers of people, and in the poorer countries of the world. However, the search for genes is problematic when it moves across the threshold from purely medical to socially significant issues, and when it is suggested that genes are responsible for a wide variety of attitudes, positions, and behaviors. Recognising the enormity of the problem caused by current research on genetics and behavior, the Nuffield Council in the U.K. has recently released a report documenting these concerns and warning that parents should not be able to know or select their children on the basis of intelligence, personality, or sexual orientation, even if there were a way of doing so (Nuffield Council on Bioethics 2002).

Hamer and Copeland (1996) have claimed that there is a genetic basis for social and psychological behaviors, including alcoholism, predisposition to violence and antisocial behavior, and homosexuality, as well as language and intelligence. Claims have even been made for genetic causes of personality types and "addictive personalities" that determine poverty and "homelessness," as if our socioeconomic status is dictated by one's genes.

The search for genes responsible for these social patterns of behavior has now been pursued in earnest for at least a decade, and there is no sign that the search parties are giving up, despite the lack of repeatable results. The lure of this research is not solely prejudice but also the dream of quick fixes. Social ills, if they are ills or ought to be perceived as such, are equated with medical diseases and often treated as if they were public health issues, as if it were just a matter of finding the antidote. Alcoholism and homelessness may be associated with medical care and housing and they may cost societies a great deal of money, but they are not like physical diseases; there is not just one single biological cause that can be identified and then defeated. From the point of view of genetics, it is most unlikely that any unambiguous answers to social issues such as these, let alone magic cures, will ever be forthcoming.

GAYNESS AND THE MEDICAL MODEL

If we move from the paradigm of gene searches to find "cures" for public ills to the search for gay genes, we have to ask how this fits, in any way, with the models to which gene research has been applied. As we know today, sexual orientation does not belong to the medical model, and it is not a public health issue—homosexuals do not fill more hospital beds than heterosexuals. This is true even when including HIV/AIDS, which ravages high numbers in heterosexual populations worldwide and at higher proportions than it does homosexuals. Homosexuality is not imparted by a virus; does not arise in only one specific social milieu; and is independent of educational attainment, state of health, intelligence, income, or any other social variable. Homosexuality is not a marker of a poor gender identity either (Ross 1983a). Indeed, sexual orientation and gender identity are separate to the extent that one does not predetermine the other. Moreover, there are identical twins who do not share the same sexual orientation, even if they have grown up together (Eckert et al. 1986), suggesting that the "genetic code" for homosexuality may well not exist. In fact, a study reporting that the incidence of both identical twins being gay or lesbian was higher than expected by chance failed to note that half of the identical twins did not have the same sexual orientation (Bailey and Pillard 1991). So why is there a fanaticism to discover "the" cause of homosexuality, and why have such large sums of money been made available over the years to account for it, particularly since heterosexuality has never been explained scientifically in the ways now being sought for homosexuality?

Homosexuality is certainly one of the categories of human behavior with an extraordinary history. Condoned, ritualized, or even admired in some cultures (Herdt 1993), homosexuality in the Judeo-Christian tradition was considered to be a sin. In the modern era, "sin" has translated into deviance and deviance into criminal records and psychiatric designations (Rowse 1983). Until the 1830s, the English word *normal* meant standing at a right angle to the ground but then, increasingly, the term began to designate conformity to a common type. Similarly, in France, it took on this connotation only in the 1840s when Auguste Comte spoke of normal states of organisms. This places the modern meaning of "normal" and "normality" at the onset of the industrial revolution. Ivan Illich (1977) argued that the term has served advanced industrial societies extremely well. At the very moment that universal standards are defined, established, and then applied, anything falling outside those standards can be defined as abnormal and, conceivably, can be legitimately attacked (see also Weeks 1981). In nineteenth-century England, homosexuality was punished by hanging. In the twentieth century, under Hitler, homosexuals were sent to concentration camps and killed, and, under Mussolini, celibacy and being an unmarried adult was punishable by a prison sentence (Kaplan 1992). There are a number of cases in the 1960s and 1970s of psychosurgery (frontal lobotomies or leucotomies, cingulotrachotomies and hypothalamic lesions) being used to "treat" and/or "cure" homosexuality. Chemical castration by application of antiandrogen therapy has also been used widely. A similar effect is achieved by estrogen therapy. Aversion therapy has been used in many Western societies. Among the common punishments used in aversion therapies are the administration of electric shocks or emetic agents that cause bouts of prolonged vomiting (Rowe 1962). An extreme form was hot-plate therapy, asking the subject to place his hand on a hot plate for punishment if he dwelt too long on the "wrong" picture (Watson 1979). This has more to do with punishing "deviant" behavior than modifying hypothesized biological states that have not been shown to exist. These "treatments," one presumes, were permitted in order to create a society that Delgado (1971) once termed the "psychocivilised society."

In cases of lobotomy, and any surgical intervention causing brain lesions, the lines became very blurred between medical need and social control. As an aside, surgical intervention extended also to controlling violence. At least this was high on the agenda in the United States of the late 1960s (Mark and Ervin 1970). Today, methods for such control "are a little more sophisticated . . . and drugs rather than the knife become the approved approach" (Rose 1995, p. 381; cf. Kramer 1994). Among these "more sophisticated methods," which Rose condemns, are those directed at lesbians and homosexual men.

Whether homosexuality is seen as a crime (deviant behavior) or, in psychiatric terms, as abnormal, it is overladen with negative meanings and, in many contexts, has been considered taboo. It was less than thirty years ago that homosexuality was eliminated from psychiatric handbooks as a mental disorder. According to the law books and the psychiatric handbooks, homosexuality has now been freed from the burden of abnormality in Western countries. Yet one cannot avoid noting that contemporary Western research aimed at finding evidence for genetic causes of homosexuality, and sometimes lesbianism is motivated by the perception that these behavior patterns are problems, if not deviant. Consequently, we must recognize that this sort of scientific research is not value-free and objective, but rather it is undertaken with a social application in mind.

Some apparently well-meaning parents-to-be might argue that being homosexual would make their child's life difficult and, therefore, they would prefer to prevent giving birth to a homosexual child as some sort of act of love. New terms have been created identifying a "procreative beneficence" and encouraging us to see reasons why we should select the best children (Savulescu 2001), thus presumably morally legitimizing and guiding genetic research in the direction of a new eugenics. The aim is to identify the genes responsible for specific behaviors and then advise individuals who carry them not to reproduce, thus ultimately fulfilling the marketing promise of producing only "perfect" babies and presumably what is considered to be a "healthy" society through genetic engineering. Of course, the possibility of compulsory sterilization of those carrying the identified genes also lurks in the wings: we should not forget that Canada, Sweden, and Switzerland continued to practice sterilization on eugenic grounds until at least the 1960s and in Peru it continues today at a massive rate (Nuffield Council on Bioethics 2002). And what will happen to those slipping through the net and developing the "symptoms" that were to be eradicated? They might face stringent social victimization. By attempting to evict "problems" the planners have not thought about what will happen to the individuals identified with any of the target problems. "Behind the rhetoric of universal rights," said Evans, "there stands a citizenship machinery which effectively invades and corrals those who by various relative status shortcomings are deemed to be less than fully qualified citizens" (1993, p. 5).

SCIENTIFIC/MEDICAL STUDIES OF HOMOSEXUALITY

In the second half of the twentieth century, "scientific" writings on homosexuality were rekindled by Alfred Kinsey's findings (Kinsey et al. 1948a,b). Kinsey's findings showed to an astounded audience that a very high percentage of males and a much higher than expected percentage of females had engaged in homosexual activity at one time or other. This was morally and socially so unacceptable that studies were undertaken to show that homosexuality was not the behavior of "normal," well-adapted people but the consequence of a biological abnormality or a congenital defect (although the evidence should have invited some to regard it as far less of an aberration than it was seen). The spate of publications to follow seemed to have as their main aim to disprove Kinsey and to tighten up moral concerns about such claimed widespread practice of homosexuality.

Dickinson (1948) concerned himself with lesbians and measured the erectility of breast and nipples, the labia majora and minora, pubic hair, and clitoris erectility and size. One needs to note that, as in the nineteenth century, measurement of physical features became the guiding procedure for establishing difference (i.e., inferiority or abnormality). However, now the customary measurement of brain size had been dropped in favor of measuring the size of genitalia. From these measurements of a total of 31 women Dickinson claimed to have found significant difference in the size and "behavior" of the sexual organs of lesbians in comparison to, as he calls it, his experience in "office practice"! Of course, we do not see any data of his daily experiences, that is, his office practice. Variation of anatomical difference on an extremely small sample was thus read as abnormality for all lesbians, and served to confirm the medical model of homosexuality as a disease (Henry 1948). There is no evidence that biological variation in genital construction (size of penis or of clitoris) or any part of the anatomy is intrinsically linked to behaviors such as homosexuality or, indeed, transvestism. There is, in fact, evidence that the latter is based on conditioning rather than biology (Walters and Ross 1986).

Hormonal causes for homosexuality were considered seriously, as they still are in some quarters today. The writings of D. J. West, a psychiatrist and reader in clinical criminology at Cambridge University, were influential. His book, first published under the title *Homosexuality* in 1955, went through four editions and seven reprints between 1955 and 1977, and then was revised as *Homosexuality Re-Examined* in 1977. Even in the revised version he argues that "the possibility that homosexual behavior in humans is caused by some glandular deficiency cannot be dismissed out of hand" (p. 65). According to West, androgen levels influence the strength of sexual desire, and an excess of prepubertal androgen in girls "may masculinize certain aspects of their social attitude and temperament." He quotes lowered sperm count and relative infertility as another relationship between male homosexuality and biology. Of course, West's reductionist position was strengthened rather than weakened by the later works of scientists like Money and Ehrhardt, and Dörner in the 1970s and 1980s, who supported the medical view of homosexuality as an abnormality induced either by certain oversupplies or undersupplies of relevant hormones at a crucial stage of prenatal development. We may also note here that infantility and infertility as a result of general underdevelopment (immaturity) were arguments proposed for homosexuality, very much in the same manner as they were earlier for women and so-called nonwhites (see Chapter 2).

Geschwind and Galaburda (1987) took up the hormonal theory of Money (see Money and Dalery 1976) and modified it to hypothesize that male homosexuality may be caused by the action of testosterone on the development of the left hemisphere of the brain. In addition, Geschwind and Galaburda (1987) were prepared to extend their hypothesis to offer some biologically determined associations of homosexuality. Here they displayed the often tortuous path of reasoning and tenuous assumptions so frequently used for biological explanations of difference. They were keen to explain some anecdotal (and definitely not proven) indication that homosexuals have a higher rate of "non-righthandedness," meaning left handers plus those with no significant hand preference. First they adopted the former study by Dörner and colleagues (1983), claiming to show that stress of mothers during pregnancy leads to increased levels of homosexuality in their male children (we discussed problems with his interpretation of the results of the study in Chapter 7) and linked it to research showing that stress affects levels of testosterone. Stress initially causes testosterone levels in the male fetus to rise and later, after birth, results in permanently lower levels of this hormone. By extrapolation of these results, Geschwind and Galaburda concluded that stress in pregnancy in humans raises the level of testosterone in the male fetus and so disrupts the development of the left hemisphere, leading to more "non-righthandedness," and also homosexuality. Therefore, they argued, a greater proportion of homosexuals should be left-handed. Then they were prepared to go one step further. An elevated level of testosterone in the fetus also suppresses the development of the immune system. Thus, homosexuals may have impaired immune systems and, they argued, this may be why they are more susceptible to AIDS.

By adopting a unitary, biological cause for homosexuality they were able, in one neat parcel, to suggest an explanation for the epidemiology of AIDS; thus reductionist explanations subsume other sciences and social sciences. Although the male sex hormone may influence the development of the brain, there is no evidence that it does so in the manner they suggest, or furthermore that this might have anything to do with causing either homosexuality or left-handedness or non-righthandedness (Vines 1992). In fact, there is no consistent evidence for a greater incidence of left-handedness in homosexual men or for unusual levels of the sex hormones (reviewed by Byne and Parsons 1993).

In the same vein, Cheryl McCormick and colleagues (1990) interviewed 32 lesbians and found that the majority had some degree of left-hand preference. This convinced her that lesbians have an atypical pattern of specialization of the left and right hemispheres and therefore that there is a neurobiological difference between homosexual and heterosexual women. From a left-handed preference she too extrapolated to hormonal levels in the body, arguing that this difference was due to lesbians being exposed before birth to an *excess* of testosterone. One notes that at least 35 percent of the general population show some degree of left-handedness, presumably not all because they are exposed to abnormal levels of testosterone. Obviously, such facts considerably weaken the foundation of this hypothesis. There is no substantial evidence to support the notion that lesbians have been exposed to a different hormonal environment before birth, although some researchers were keenly pushing this view for over two decades (e.g., Dörner 1976a,b, Green 1993) and continue to do so today (Green 2002). This is yet another example of grand theory building based on assumptions that have already been shown to be incorrect.

It is noteworthy that in every theory stipulating a hormonal cause of homosexuality, it is assumed that homosexual men are more like women than heterosexual men, and that lesbians are more like men than heterosexual women. This thinking is based on traditional notions about homosexuality, mirroring assumptions of heterosexual opposites that have long since been shown by psychosocial research to be incorrect (see Byne and Parsons [1993] for a more complete discussion of this point).

In fact, research on sexual orientation is never far removed from research on gender because gay men and lesbians are considered to fall outside the norms for their particular sex, and homosexuality is often, quite incorrectly, said to be associated with adopting the patterns of thinking and behavior characteristic of the opposite sex (e.g., Phillips and Over 1995, Pillard and Bailey 1998). We might therefore predict that the outcomes of research on homosexuality would affect thinking on the differences between women and men. If sexual preference were found to have its roots in an individual's genes or hormones, there would be yet another buttress to support the dividing wall between women and men. We are convinced that the concept of the hypothetical "gay gene" or the "gay mix" of hormones will be extended beyond the interest in sexual preference to influence thinking on gender.

Another example of such thinking and tying it to a biology foundation is the research of LeVay (1991, 1993) claiming to show that homosexuality might be associated with just one area of the brain, and a region of the brain not involving higher and more complex levels of processing. LeVay measured the sizes of four nuclei (collections of nerve cells) in a lower part of the brain called the hypothalamus and did so by sampling post-mortem brains of gay men, heterosexual men, and a group of women of whom he had no recorded sexual preference.

It would normally be a very difficult task to get a sufficient sample size of gay men's brains to make a comparison worthwhile but LeVay was able to obtain the material he needed from gay men who had died of AIDS. We have to note here that, although some of the heterosexual group in his sample had died of AIDS, the proportion was much smaller than in the homosexual group. Although LeVay did not discuss this, it presents a problem because viral infections of the brain can be concomitant with AIDS and might affect the size of nuclei in the brain, even quite specifically. Setting this lack of control aside, LeVay found evidence that the average size of one of the nuclei, labeled INAH3, was smaller in the homosexual men than in the heterosexual men and more like that of the women in his study. Although LeVay used a sample of only six women, all of whom he presumed to be heterosexual although he had no reason to do so, it had been known before that INAH3 is smaller in the brains of women than in men. In fact, this is why LeVay chose to look at this region. Also, in studies using animals, primarily rats, a region of the hypothalamus called the medial preoptic area has been shown to regulate the level of sexual activity and control some of the behavior patterns involved in sexual intercourse (particularly those typical of males) but not choice of a sexual partner.

Could such a small area of the hypothalamus control the complex decisions involved in choice of a sexual partner? Is that where a hypothetical "gay gene" is acting? It is little wonder that the findings of LeVay received much media coverage. Although LeVay (1993) mentioned that experience could affect the size of brain regions, he made it clear that he is of the opinion that the scientific evidence points to a strong influence of nature, and only a modest influence of nurture.

But this interpretation of the result assumes that the cause of a smaller INAH3 is linked to a gene (or genes), perhaps to hormones and then to making the nucleus smaller and so, finally, causing homosexual behavior. It is equally probable that behaving as a homosexual may lead to a smaller size of the nucleus, perhaps by way of hormones. In fact, recent evidence shows that the levels of sex hormones circulating in the blood of adults (rats in the study) can change the size of nuclei in the brain (Cooke et al. 1999). We are not suggesting that this is known to be the case for INAH3 but we want to emphasize that merely finding a difference in the size of the nucleus gives no indication of the cause. Added to this, it is most unlikely that the complex behavior of partner choice in humans is controlled solely by one nucleus in the lower part of the brain. Higher cognitive processes at the cortical level are very likely to be involved.

Another study reported that the anterior commissure (a tract that connects the left and right sides of the brain) is 34 percent larger in homosexual men compared to heterosexual men (Allen and Gorski 1992). Again, almost all of the brains from a homosexual group came from patients who had died of AIDS, whereas very few of the heterosexual group died from this cause. This illustrates again that claims being reported in the scientific literature are not based on properly controlled studies, a fact that is lost when the information is picked up and propagated by the media. Allen and Gorski (1992) also jumped to an assumed causation from hormones to size of the brain region to homosexual behavior. They drew a direct link between the action of the sex hormones during development and the claimed difference in the size of the anterior commissure. In their view, the hormones have a global action on many brain regions, including the higher centers to which the commissure connects, and the differences in all of these regions cause homosexuality. In fact, their results have been the basis for postulating a general constellation of cognitive differences between homosexuals and heterosexuals, a field being actively pursued by some researchers (see Hines 1993). Given that there may be differences in lifestyle between the majority of homosexuals and heterosexuals, there may well also be differences in cognitive function, but that does not imply that these differences are caused by genes or hormones any more than it implies that lifestyle determines cognitive function.

Yet another study has examined the size of a nucleus in the brains of homosexual and heterosexual men, this time the suprachiasmatic nucleus, another small region of the hypothalamus (Swaab and Hofman 1990, 1995). It was found to be larger in the sample of ten homosexual men than in the sample of six heterosexual men. This time the samples were controlled by comparing both heterosexual and homosexual postmortem samples from people who had died of AIDS. The researchers were not so rigorous in the interpretation of their result. Despite mentioning that experience as well as genetic and hormonal influences could have caused the size difference in this small region of the brain, Swaab and Hofman were inclined toward the notion that this region causes differences in sexual orientation and associated behaviors. They were less partial to the possibility that behaving as a homosexual could have caused the difference in the size of this region of the brain. They even extended their results to an explanation of sex differences in behavior and suggesting that functional sex differences in reproduction, gender, and sexual orientation could stem from anatomical differences in the hypothalamus. In addition, and based on evidence from animals that this particular nucleus is known to control daily cycles of activity, they thought that the size difference in the suprachiasmatic nucleus might cause different sleep-wake cycles in homosexual versus heterosexual men, and drew on some dubious information about urban gay lifestyles to "prove" their point.

These examples are but part of the effort to study homosexuals from almost all aspects of biology. Today they are the focus of molecular geneticists in search of the "gay gene" or "gay genes," as we will discuss later.

FROM TREATMENT TO PROTEST

In the years following World War II the punitive legal procedure against homosexuals was changed into an invasive, even mutilating, "treatment" model. Part of the issue in the 1950s and 1960s was considering homosexuality a perversion within internationally defined disease models. It was thus a medical problem, treatable with medical tools or explicable as an abnormality of brain function.

Behind the "treatments" and the search for physical and physiological abnormality hid the profound oppression of a substantial number of people. Negative attitudes died hard, even in the latter half of the twentieth century. In the United States alone, countless studies have demonstrated an insipient persistence of negative stereotypes and attitudes. For instance, a study of doctors (Matthews et al. 1986) as well as nurses revealed overwhelmingly negative views of lesbians (Young 1988). In Young's study, 64 percent of interviewees felt pity, disgust, repulsion, fear, or at least unease and embarrassment in facing a known lesbian. These opinions were expressed with regard to colleagues and patients. In another study (Randall 1989), involving nurse educators in the degree of nursing studies, 52 percent believed lesbianism to be unnatural, immoral (23 percent), illegal (19 percent), a disease (17 percent), and/or perverted (15 percent). They thought that lesbians transmit AIDS (20 percent), molest children (17 percent), and are unfit to work as nurses (8 percent). They further avoided lesbian issues in the classroom (54 percent) and expressed being uncomfortable teaching or providing care to lesbians (28 percent). In the context of constant and powerful onslaughts by such public institutions as the medical profession, science, the police, the church, and the law (Thompson 1985), it was rather difficult for many gay men or lesbians to think of themselves outside those deterministic constructs of defects, deviation, and decadence.

Plummer (1999), in his detailed study of homophobia, has found that derogatory words such as "poofter" and "faggot" are used by boys often long before puberty, long before their sexual identity is formed, and before they know anything concrete about sexuality. He showed that these terms are deeply negative and used in preteenagers for a vast array of nonsexual behaviors (also discussed by Corbett 2001). His own Australian study confirmed that children learn very early in their schooling to call other students of their grades "poofters," often not even because of specific behaviors or suspected interests in other boys but because a particular choice a fellow student has made may not fit their masculine stereotype. Just about anything can become a target for name-calling if the group has created certain gender schema and has ordered all life activities accordingly. The reason for name-calling and bullying may be made on the basis of a boy's choice of an unpopular or "female" school subject, hobby interests, performance in sports and the like. Plummer notes that once these names have been used in this fashion in school the meanings and consolidated attitudes persist into adulthood.

In the last four hundred years or so, male homosexuality has always been fought more vigorously and been the subject of more overt persecution than lesbianism, partly because the latter was barely known or understood by male lawmakers and professionals. Lesbians suffered from the veil of invisibility (Roberts 1983) for a long time. This invisibility was so complete, as Kennedy and Coonan (1975) argued, that less was "known about the lesbian and less accurately than about the Newfoundland dog" (p. 34). This blind spot about same-sex love in women was, of course, deeply entrenched in the male psyche, considering it an impossibility to have sex without a penis. Yet, in different ways and for different reasons, gay relationships in males and females suffered from a similar kind of silence, oppression, and fear (Rich 1980). Silence has also at least two other psychosocial consequences. Lesbian relationships, when they came to be known, were seen as being practiced mainly by women who had missed out on a male suitor (Storr 1964), or were mad or immature (Chesler 1972, Matthews 1984). If not personally vilified, lesbians also had many social disadvantages emanating from social institutions, be this concerned with taxation, superannuation, hospitalization, death (Gentry 1993), even immigration (Hart 1992, 2002), and other situations. Discrimination was enshrined in law and suffered alike by lesbian and gay male couples.

The Stonewall riots of gays against police in New York in 1969 gave the impetus for change throughout the entire Western world, paving the way for new, alternative attitudes and for the claim that gays had the right to exist (Kaplan 1996). This right to exist and to come "out of the closet" was perceived as a deep psychological need, a cultural right, and, ultimately, as a social justice issue for modern democracies (Duberman et al. 1989). Between the 1970s and the early 1990s, homosexuality began to assert itself as a visible force, despite ongoing resistance and, if not as an alternative lifestyle, then at least as a sexual orientation worthy of equal civil rights. Admitting to gay orientation was no longer so openly victimized by the medical and legal professions and, over the remainder of the twentieth century a number of legislative changes slowly began to close some of the gaps between heterosexual and homosexual civil rights.

In the 1980s and early 1990s, queer theorists appeared on the political and cultural front, finding great risk in politically identified categories, such as gays and lesbians, to which Anderson (1983) might also have referred as "imagined communities" (Anderson 1983). Anti-essentialist in philosophical orientation, queer theorists argued that a politics of identity was of great disservice and politically dangerous because it reinforced and fed categories such as "race," gender, and sexual orientation as if these were immutable "natural" identities (Rubin 1984). Consequently, they became opposed to the civil rights agenda proposed for gays and lesbians on the grounds that it was too conservative because, as Steven Seidman (1995) argued:

> [A] theoretical and political project which aims exclusively to normalize homosexuality and to legitimate homosexuality as a social minority does not challenge a social regime which perpetuates the production of subjects and social worlds organized and regulated by the heterosexual/homosexual binary. [p. 126]

Seidman's misgivings were written with the hindsight of knowledge of the gay genes studies and these studies, indeed, lent substantial support to his views. Finding a gay gene would, in a sense, do precisely this. It would legitimize homosexuality as a genetic condition and therefore, according to the constitution, outlaw any discrimination against gay groups or individuals. However, as became very clear in response to the claims of having found a gay gene, the living homosexuals would be the last of that minority, if some had their way. Medical intervention and abortion of fetuses with the gay gene would eventually see to that.

However, radical postmodern queer theory has been grounded in individualistic and "self-transformative" perspectives to a point where basic political goals become difficult to pursue, such as equality before the law, rights to nondiscrimination in employment, and other basic human civil and social rights (Ball 2001). Another signature of the postmodernist stance, not just of queer theory, is to argue against any form of truth claims and against any universality (including being human).

Over several decades, this positioning resulted, by and large, in a rather apolitical, individualistic, and definitive anti-enlightenment stance. While contestations and resistances were expressed in writing, they were resistances that quite often looked inward instead of outward, as if the things to be contested and resisted were only part of a discourse and a text. In being purposely vague, fluid, and anticertainty, this may have even contributed to opening the doors to biological and genetic theories that now seem to offer certainty and simplicity, present so-called facts (even if these were often closer to fiction) and proffered answers instead of questions. Reductionism is an extreme counterposition of postmodernism and, it seems, we have to live with both.

While we have new names for current ideologies, theories, and countertheories, they often do not strike one as being intellectually all that new, or at least not with respect to the things that they might actually represent. One is reminded of Dostoevky's "The Grand Inquisitor" (written in the nineteenth century). In it, Christ, having returned to earth, asks the inquisitor why his thoughts of love have been perverted into submission. The grand inquisitor replies that people do not want truth and freedom, they want miracles, mystery, and authority. Postmodernists deny the existence of truth (as an absolute) and, abstractly, also the possibilities of freedom. Having deconstructed both, "authority," in the form of rather extreme right-wing politics, has effectively and successfully tiptoed into the empty spaces. Genetic determinist theories (claiming immutable categories of gender and "race," for example) sit very comfortably with this new "authority" and, one presumes, therefore have succeeded so well and are continuing to do so. Finding a "gay gene" would be a fitting tribute to a specific political milieu.

GAY GENE STUDIES

On the surface, it would be quite possible to argue that the scientific revival of a genetic basis for homosexuality in the 1990s was seen to play well into the hands of some homosexual groups. If there was a gay gene, one could not lay blame on gay people. In the United States some gay groups seized on this as a reason for claiming civil rights: if gay men and lesbians were not making a choice of lifestyle but instead were genetically determined to be the way they are, then they should be protected by the bill of rights. The problem with this stance, apart from the fact that the gay gene or genes may not exist, is that it crystallizes and reifies difference, and once that has happened, the labeled group can be singled out for negative as well as positive treatment. With a change in the political climate, gay-gene carriers (if they exist) might be singled out for persecution, as their treatment in the Holocaust reminds us.

The possibility for the existence of a gay gene arose first in the context of debates about a supposed gene for altruism in the widely read work by E. O. Wilson in the 1970s. There has been much disagreement over the existence of a gene for altruism (cf. de Lepervanche 1984). Most sociobiologists agree that no such gene could exist for long in a population because it would soon be lost in competition with the "selfish" genes. Yet, according to Wilson (1975), the gene for homosexuality, which he hypothesized to exist, must be linked to a gene for altruism or it too would have been lost from the population because, in Wilson's perception, homosexuals do not reproduce. Wilson's idea is that the gene for homosexuality persists only because homosexual individuals also carry a gene for altruism that leads them to assist in the raising of their siblings and their siblings' offspring; that is, they assist the reproduction of individuals who carry a certain percent of their own genetic material. Thus, "the gene" for homosexuality, so Wilson believed, is an aberration that persists only as a consequence of the coexistence of a protector gene, and the altruism/helping behavior determined by this protector gene is confined to individuals with whom there is shared genetic material. Altruism outside the family is biologically unsound and, if it exists at all, it will soon die out! Had Wilson turned to some research records while writing in the early 1970s, he might have been surprised to discover that a considerable percentage of homosexuals do reproduce (see also Churchill 1967, Dank 1972, Imielinski 1969, Kinsey 1948b, Ross 1972, Ross 1983b).

The idea of heritability of gayness, as of any trait, would not by itself say that such genes would be expressed in an individual. Many would rightly argue that to carry the gene would at best (or worst, depending on how homosexuality is viewed) mean only that there is a propensity or predisposition for the behavior. That is, the gene might be expressed in certain environments and not in others. Thus, if we consider for a moment the hypothetical situation in which there might be a gene for lesbianism, it would only increase the chances of individuals to be lesbian. It follows that, in certain environments, this gene might not find expression. If these environmental conditions were known, it is argued, they might be avoided, and thus "the problem" could be eliminated. People who reason in this way are not talking about using eugenics to "eliminate the problem," but rather are practicing social control by manipulating the environment (family and school life, gender identity, learning, and so on) so that the offending gene cannot be expressed. In fact, no one has a clear idea what conditions might prevent the hypothesized gay gene from being expressed, even though there are retraining programs for children whose parents develop a concern that they may be gay or lesbian. Apart from the oppressive position these represent, they are a means of making money out of parental fears based on dubious evidence.

But despite substantial media exposure and hints in the direction of affirming that there might be a genetic basis to homosexuality, no such causal link between a specific gene and expressed homosexuality has, in fact, been found, and the studies that have claimed to have done so have ultimately failed to convince the scientific community.

In 1993, a research team (Hamer et al. 1993) claimed to have isolated a gene sequence for male homosexuality. It was said to be located in a small stretch of DNA on the X chromosome. We note that this piece of DNA, which they called Xq28, contains hundreds of genes. The researchers recruited homosexual men into the study through an outpatient HIV clinic or through advertisements in homophile publications. First they interviewed the subjects about the incidence of homosexuality among their relatives. Forty pairs of homosexual brothers were selected who had homosexual relatives on their mother's side of the family but not on their father's side. The reason for this choice was that the researchers were interested in looking for sex-linked gay genes carried on the X chromosome. As a male inherits his Y chromosome from his father and his X chromosome from his mother, any characteristic carried on the X chromosome will be passed on from mother to son and not father to son. Thus, by selecting families in which homosexuality was more common on the mother's side of the family than on the father's, Hamer and colleagues were narrowing down their search for the "gay gene(s)" because they could concentrate on the X chromosome and exclude all of the others.

They chose gay men with gay brothers and gay uncles and cousins on their mother's side of the family. Gay men with gay uncles and cousins on the father's side of the family were excluded. They also excluded pairs of brothers who had more than one lesbian relative on the grounds that they were looking for a "gay gene," not a "lesbian gene." They took samples of blood from the homosexual brothers and from other family members. These samples were used to amplify the DNA and then they could identify genetic markers, which allow detection of particular sequences of genes on the chromosome. The genetic markers on the X chromosomes taken from each individual were then matched up with those of each subject's homosexual relatives. The idea behind this genetic analysis is that on average each pair of brothers will have in common about half of their DNA on the X chromosome (and all of their other chromosomes too), and if both brothers are homosexual because they inherited the same gene, the "gay" gene must be located on the parts of the X chromosome that they have in common.

Five markers at the tip of one arm of the X chromosome (Xq28 region) were found to be matched in thirty-three of the pairs of brothers. The researchers therefore claimed to have isolated a gene sequence that gives a person a predisposition to become homosexual. They concluded by saying that their study needs to be repeated and the Xq28 region of the X chromosome needs "fine mapping" to identify "a specific gene . . . where and when it is expressed and how it ultimately contributes to the development of both homosexual and heterosexual orientation" (Hamer et al. 1993, p. 326). The same research group extended its own study by reporting, in 1995, on two new series of families containing either two homosexual brothers or two lesbian sisters (Hu et al. 1995). The Xq28 sequence match was found for the families with gay males, in line with the previous result, but not for families with lesbians. This result suggested to the researchers that lesbian preference has nothing to do with the Xq28 gene sequence. Indeed, if they are correct, and it seems they are not, different factors must be involved in male homosexuality and lesbianism.

Hamer and colleagues believe that a single "gay gene" will be found in the Xq28 sequence. To these scientists the tendency to be homosexual is somehow embodied within a single gene. This hypothesis is, in our opinion, not sensible. It is a result of reductionist thinking and greatly oversimplifies the factors that interact to determine choice of a sexual partner. Is the "gay gene" expected to exert its effects in the cells of the brain? Some researchers think so, as made clear by Pillard and Bailey (1995): "The gene(s) in question is presumably expressed in the brain but finding it and learning what it does will require considerable further effort" (p. 79).

The paper reporting the findings of Hamer and colleagues stirred immediate controversy and the approval was short lived. The Research News section of *Science* may have waxed with excitement over the Xq28 discovery, stating that the research was better controlled than former studies claiming to have found genes for schizophrenia, alcoholism, and manic depression, but the author of this article (Pool 1993) seemed to base most of his enthusiasm on the fact that Hamer was considered to be a good solid scientist with a reputation for unraveling genetic codes in yeast. There was no questioning of his research at the sociological level of the study. When the Hamer study was published in 1993, the authors tried to encourage other research groups to repeat their work. As we have mentioned, the same research group extended their own study by reporting, in 1995, on two new series of families containing either two homosexual brothers or two lesbian sisters. No other research groups have replicated their results. An attempted repeat by Rice and colleagues in 1995 failed to replicate the findings, and some technical issues brought about confidential investigations of Hamer's research by the Office of Research Integrity (ORI) in the same year (Gonsiorek 1995, Marshall 1995). In 1999, Rice and colleagues published a report in *Science* that repudiated the results of Hamer's research group. They flatly said that their results were very discrepant from Hamer's and did not support an X-linked gene underlying male homosexuality, thus confirming the growing doubt about the entire theory of a gay gene (Wickelgren 1999).

Let us go back to the early 1990s. In the space of three years (1991–1993) five major reports were published linking homosexuality to genes, as we mentioned (Allen and Gorski 1992, Bailey et al. 1993, Bailey and Pillard 1991, LeVay 1991, 1993) followed by the Hamer and colleagues study of 1993, their further study in 1995, as well as their book (Hamer and Copeland 1996). Although there were a number of very good critiques published at the time (e.g., Hubbard and Wald 1993), these received far less attention than the research claiming to have found a gay gene. An excellent summary of the various responses to the debates in the United States and Britain showed that the gay gene story, especially by Hamer, made headlines and retained front-page news status for some time (Conrad and Markens 2001). Remarkably, public responses largely fell into two rubrics: a "rhetoric of concern"(British Press) and a "rhetoric of promise" (American media), but often with little concrete evidence of knowledge or debate on the scientific basis for the research reports by Hamer and colleagues (Conrad and Markens 2001). A rhetoric of concern followed the line that one had to view the team's efforts with skepticism, and if the hypothesis was true, that one was rather more concerned than pleased because of the social consequences this could produce. A rhetoric of promise in the United States, on the other hand, greeted the "findings" as another great breakthrough.

Yet, while the results of Hamer's gay gene research have thus far not been sustainable, the social construct of the "gay gene" has persisted (Conrad and Markens 2001). In some areas in the United States the finding met with positive attitudes within the gay community, especially in those states with laws that proscribe against victimization of individuals on the basis of genetic difference (Henry 1993). Other gay groups welcomed the news of an alleged genetic cause for homosexuality because it removed the concept of choice (Carr 1993) and their sexual orientation was thus ultimately not their fault. Elsewhere, however, the potentially sinister uses of this information raised deep concerns (Vines 1992). For instance, the gay gene media coverage produced a flood of letters of inquiry from homophobic couples wanting to know how early one could take tests during pregnancies so that a fetus afflicted with this "gay gene" could be aborted within the legally permitted time frame.

One quite commonly encounters people now who speak of "gay genes" or a single "gay gene" and even a "gay brain," as if the genes themselves, or the brain itself, were somehow gay. It is as if the genetic material holds the blueprint for gayness and makes its stamp on the brain. Being gay in this way becomes something genetically abnormal, a flaw in the ground plan. There has been no attempt to find the genes for heterosexuality, rather than homosexuality. The search for genes more often than not begins with defining a characteristic as abnormal. Socially, this is a poor starting point, let alone a basis for any scientific investigation.

How valid is the claim for a gene or genes for homosexuality? As for every aspect of behavior, there are likely to be both biological and environmental factors that contribute to sexual orientation, a characteristic that may vary throughout an individual's life span. We do not know what these factors are, but we do know that they are likely to interact in extremely complex ways (see Ward 1992), and it is important to ascertain the scientific accuracy of such claims.

With hindsight, having a study in hand that has failed to replicate the results by Hamer and his colleagues, one begins to wonder how and why curiosity to find a cause (and preferably a single cause) for homosexuality has remained so very high. No single study has been so engaged in spending high research dollars to find a "cause" for heterosexuality.

METHODOLOGICAL FLAWS IN STUDIES ABOUT GAY POPULATIONS

Few studies searching for a biological cause of homosexuality bother to define what they mean by the term *homosexual* or give a full description of the volunteers who take part in the study. A definition of the homosexual used in a study investigating homosexuals would appear to be essential. When is a person gay or lesbian? How many same-sex sexual encounters would a person need to have had to be classed as gay or lesbian for the purpose of the study? Would the scientist consider choosing a sample from more than half of the male North American population that Kinsey's surveys found to have had a homosexual experience reaching orgasm at least once in their lives, or from the quarter of females who have had at least one lesbian relationship? Or is it the "self-confessed" homosexual who, for political or other reasons, wishes to be identified as such? Queer theorists would be appalled, but perhaps not surprised, to find that scientists based their research on assumptions of a stable homosexual category.

As some surveys have shown, a majority of individuals in a society have taken part in some form of homosexual activity at some time during their lives. Hence, the distinction between heterosexual and homosexual is not as clear-cut as one might imagine. Although homosexual behavior in many young people might be seen as experimental and not the same as "genuine" homosexuality, which persists into later life, this may also be said of heterosexual behavior. Homosexual experimentation may also take place at any age. The labels "homosexual" and "heterosexual" may make the differences appear more dramatic than they are in practice. There is ample evidence that the lifetime of an individual may involve homosexual and heterosexual periods or even concurrent homosexual and heterosexual relationships. The label of homosexuality is therefore more accurately applied to sexual practice rather than to an individual as a statement of identity although, for a core of exclusively homosexual people, gay, queer, or lesbian identity is a strong, and often political, expression of practice. Many gay men and lesbian women are married and have children. Many who have been exclusively homosexual through their lives, on the other hand, may not have volunteered to take part in such studies, and some who are identified as gay by their own choice actually do not sexually practice their professed orientation. So will the "real homosexual" please stand up?

Individuals who volunteer to be scientifically researched in such studies may well form a skewed subpopulation of those whom they are taken to represent, be they homosexual or heterosexual. For example, subjects taking part in the study may be from a particular social club or social class. Quite often, recruitment of homosexual subjects happens exclusively by means of gay media and gay club advertisements. This is a seriously skewed subpopulation. It is not exactly known how many practicing gay people in the population at large do not belong to a club and do not read the gay papers but they are likely to form a substantial percentage. One can assume also that many of today's club members are young city dwellers since there are few such outlets in rural areas. In cities young people, rather than older ones, tend to frequent clubs, partly in search of partners and a good time. Therefore, one cannot be sure whether a study has exposed aspects of the subjects' sexual preference and not aspects of their personalities and/or past experiences that led them to volunteer for experimentation.

Rarely are the homosexual and heterosexual volunteers matched for class, lifestyle, personality, age, past experience, or even for specific medical conditions or health impairing activities (drug or alcohol consumption). These factors are ignored at the researcher's peril. Sociologists have been at pains to uncover differences according to various human groups (such as class, ethnicity, gender, etc.) so as to sensitize us to the fact that not all humans in a given society move in the world in the same way, see the world around them with the same eyes, or do the same things. When Bourdieu (1977) argued for a methodological objectivism, he meant that this would not only extract meanings that the researchers thought were there and only needed to expose but would also uncover meanings of which the researchers knew nothing beforehand but had helped to create through the research process. Failing to take such vital basic criteria into account introduces a host of uncontrolled variables, and the results must necessarily be flawed.

Methodologically, the very first step of the research—obtaining the study samples—has been seriously compromised in many studies of homosexuality, especially those undertaken by scientists who have not learned to appreciate that social methodologies need to be as rigorous as scientific ones and that such rigors have good scholarly reasons. Indeed, Poincaré, the famous mathematician, wondered why the social sciences always talk about their methods while scientists talk about their results. But method, in science, tends to refer to experimental control, replicability, and verifiability of a piece of research. It is a failing of an undergraduate student not to consider experimental control. How much more do we need to chide researchers who publish material without revealing the details of the experimental design of the sampling method and the conduct and details of the interviews, and then expect their results to be taken seriously? We need to be able to replicate the study and verify the results.

In addition, it is not difficult to see that "volunteering" may have meant something quite different to homosexuals than to heterosexuals, especially when these studies were (and sometimes are) conducted in an atmosphere of opposition to their sexual practices and heterosexuals make up the so-called control group. Further, many scientific studies on biological differences of specific groups use interviews as a social variable but then do not present details of the interviewing technique used. It has long since been established (Grant et al. 1987) that interview techniques are influenced by gender, hierarchy, and a host of circumstances. Bourdieu (1977) rightly warned that in one's methodology one may slip from the model of reality to the reality of a model, meaning that the reality one wished to extract has actually been defined by the researcher's bias: the model becomes real (e.g., the "real" homosexual as an absolute/universal category).

The study by Hamer and colleagues can thus be criticized on statistical and technical grounds. Details of the techniques used in sampling and in the interviews were not mentioned in the paper. There were also some concrete statistical problems. For one, the researchers based their calculations on the assumption of a 2 percent prevalence of homosexuality in the general population. This is a very low estimate, even assuming there is such a discrete group. Bower (1993) pointed out that it is more generally accepted that the prevalence of male homosexuality ranges from 4 to 10 percent. According to Evan Balaban and Anne Fausto-Sterling, if the data are recalculated using the 4 percent figure, some of the claimed findings lose their statistical significance (summarized by Miller 1993). They have also expressed concern that DNA samples were taken from fewer than half of the mothers in the linkage study and the rest of the calculations were based on estimates. As the Xq28 sequence is on the tail end of the X chromosome, this problem is heightened: there are inaccuracies in tracking down markers next to a sequence in this location.

The select group of men featured in the research of Hamer and colleagues represents only one small section of all homosexual men—those having a known or openly gay relative. Also, all of the subjects were openly gay themselves. That separates them from the very large population of gay men who prefer not to disclose their homosexuality for fear of reprisals or other repercussions. Moreover, the researchers point out that even within the selected population they studied, the Xq28 region was neither necessary nor sufficient for homosexuality to be expressed, meaning that there is no one-to-one relationship between being gay and having the Xq28 gene sequence.

So even if the results are supported by further studies, it is possible that the Xq28 sequence coded for some other aspect of behavior that characterized the volunteer group of gay men. Perhaps, as Hamer and colleagues also acknowledged, the sequence contains some genes that might lead to a more extroverted personality, although we would be just as reluctant to argue for the existence of genes that influence such aspects of personality as we would for sexual orientation. This example is raised to demonstrate the complexities of interpreting a result.

STUDY OF TWINS

There is a tradition in genetic studies to use identical twins, particularly those who had been raised apart. The study of twins has fascinated scientists for a very long time. Many of them consider that it is possible to determine the amount of genetic control of traits, be they anatomical, physiological, or behavioral, by assessing the degree of correspondence of these traits between genetically identical twins compared to fraternal twins (developing from separate fertilized eggs and thus having different sets of genes). If, for example, identical twins separated at birth share particular behavioral characteristics, despite the assumed differences between the environments in which they have been raised, it is said that these similar characteristics or traits have been inherited. Alternatively, if it is not possible to obtain a sufficient sample size of identical twins separated at birth, researchers may compare the commonality of a trait between identical twins raised together versus nonidentical twins raised together. If there is greater coincidence between the identical twins than the nonidentical twins, the trait is considered to be genetically determined.

Prior to the 1970s, much emphasis was placed on twin studies and the inheritance of intelligence, as measured on IQ tests. These fell out of fashion when Leon Kamin (1977) effectively discredited the work of their main protagonist, Cyril Burt, a British psychologist. Kamin showed that Burt had overstated his case for the genetic inheritance of IQ, and that he had even invented much of his data. Furthermore, it was never proven that twins separated at birth were in fact raised in entirely different environments. This criticism is particularly salient because, when twins have to be separated, most adoption agencies seek strenuously to place each one in a family of social and economic status equal to their family of procreation. These considerations led to a waning of interest in twin studies until the recent resurgence of interest that has occurred along with the development of new techniques for studying the human genome.

A study by Bailey and Pillard (1991) looked at the sexual orientation of brothers who were identical twins, nonidentical twins, or adoptive siblings, one of the pair being identified as homosexual. The homosexual subjects responded to an advertisement and so "volunteered" for the study. Each was questioned about the sexual orientation of his brother, and in *some cases* this information was checked by questioning the brother directly. There were between 50 and 60 individuals in each group. The incidence of homosexuality in the brothers of the identical twins was 52 percent, of nonidentical twins 22 percent, and of adoptive brothers 11 percent.

The same researchers have recently conducted a similar study of lesbians, reporting the incidence of lesbian lifestyle among the sisters of identical twins to be 48 percent, of nonidentical twins 16 percent, and of adoptive sisters 6 percent (Bailey et al. 1993). Here their sample sizes were 71 pairs of identical twins, 37 pairs of nonidentical twins, and 35 pairs of adoptive sisters, all of whom entered the study in response to advertisements placed in lesbian-oriented publications in the United States. Again, the sexual orientation of the relatives was assessed by asking the volunteers to report, and, where possible, checking this with the relatives. The results were interpreted as evidence for genetic inheritance of a lesbian orientation because the concordance of this behavior was higher in the identical twins than in the other two groups. This deduction, of course, assumes that any of the unknown environmental influences on this behavior are acting equally in all three groups. But it is reasonable to suggest that identical twins are treated more similarly than nonidentical twins or adoptive siblings, and that this could be the explanation for the greater concordance of their lesbian choice. Other scientists have criticized this study and ones like it for the same reason (Lidz 1993).

Since the identical twins in this study were raised together, similar environmental influences rather than genetic determinants could have caused them to adopt a similar sexual orientation. Although the authors considered this possibility, they discounted it on the grounds that identical twins when treated similarly (e.g., dressed alike) had no greater similarity in IQ scores than those not treated similarly. One might reply by saying that being dressed alike may not be an environmental factor relevant to homosexual or heterosexual choice (or for that matter for IQ performance) and that since we have no idea what diverse environmental factors shape our sexual preferences, it is naive to claim that these factors might not have been operating differently in the identical versus nonidentical or adoptive siblings.

As far as we know, there has been only one study of lesbianism in twins reared apart, and although this was limited to only four pairs of identical twins, none of the four pairs showed concordant sexual preference (Eckert et al. 1986). As we mentioned before, this result suggests that genetic factors may have no role to play in determining sexual orientation, but the sample size is too low to be sure of this. Bailey and colleagues (1993) discounted the study entirely on the basis of the low sample size, but then their motivation to find a genetic cause for lesbianism was clearly strong, as shown by their statements: "The dearth of genetic data on females is unfortunate. . . ." (p. 217) and "an elaboration of the nature of the genetic variance could be an important step to unraveling the origins of female sexual orientation. . . . These studies were designed to detect heritable variation, and, if it was present, to counter the prevalent belief that sexual orientation is largely the product of family interactions and the social environment" (p. 222).

In the concluding paragraphs of the paper, after they have fully elaborated their case for the heritability of lesbianism, the authors concede that "environment also must be of considerable importance" (p. 222). In fact, they are convinced that genetic studies such as theirs can be used to illustrate "the *nature* (our italics) of environmental influences." Is this not circular reasoning? They state that those identical twins who do not share the same orientation must do so only because the environmental influences have differed! Thus, from the start the assumption is that lesbian choice is determined genetically, and thus, twins that contradict this by displaying nonconcordance of sexual orientation must display the effects of environmental influences.

The authors end their paper by stating that these environmental factors (which suppress the expression of the gene(s) for lesbian behavior) are worth further study. We wonder why. Is it to find out how lesbian behavior can be prevented? Apparently some of the subjects of this study subsequently expressed doubts about its validity (see Horgan 1993). This indicates that some of the interviewing techniques were either not considered appropriate, or that some of the volunteers for the study may have had preconceived notions about its potential outcome and were then dissatisfied with the results. If the latter is the case, it may hint of biased sampling of the homosexual population, depending on how many of the subjects feel this way.

There is unquestionably much to criticize in the design and interpretation of this and other twin studies. Nevertheless, the search for gay and lesbian genes continues, particularly using pedigree and linkage analysis to map the gene(s) on the chromosome(s). This involves use of DNA hybridization techniques. True to most studies is a justified criticism that, at sociological/psychological levels, there is an astounding paucity of controlled methodology, not uncommon when biologists address social or psychological issues. Applications of new scientific technologies seem to give the work an immediate validity that subsumes the need for adequate controls and sufficient caution in interpreting the data.

STUDY OF FINGERS

Fingers and fingerprints have become an attraction to some researchers interested in sex differences and homosexuality. A recent study (Williams et al. 2000) made associations between the relative lengths of the second and fourth fingers (index and ring fingers), exposure of the fetus to androgen, and homosexuality. Apparently, the lengths of these two fingers are about equal in women, whereas the second finger is usually shorter than the fourth in men. The researchers extrapolated from this observed sex difference to saying that it must be caused in the fetus by the action of differing levels of androgens. Then, on the basis of Dörner's (1976a,b) earlier claims about androgen exposure of the fetus causing homosexuality, they predicted that the relative finger length should predict whether a person is homosexual or heterosexual. For example, a woman with the second finger shorter than the fourth should be lesbian because she has been masculinized before birth. The biased view of lesbianism is obvious here.

The team of researchers went out to street fairs in San Francisco, taking a portable photocopier and some questionnaires about sexual preference. Their sample of 720 people found the relative finger lengths of lesbians matched their prediction, and the conclusion reached was that maternal levels of testosterone masculinized the female fetus and caused it to become a lesbian. The paper reporting these results in the journal *Nature* gave no information about how the fingers were measured. This is not such a simple thing to do as it sounds: as the reader will discover by examining his or her own fingers, the relative lengths appear to change depending on the orientation of the fingers to the palm of the hand and to the connecting skin at the base of the fingers.

The same study found that there was no significant difference between male heterosexuals and homosexuals in terms of the relative lengths of their fingers. However, there was an effect of birth order. Men with two or more older brothers were found to have more "masculinized" finger lengths, from which the researchers concluded that androgen levels that are contributed to the fetus by the mother increase with subsequent births of male children. Now they drew on some evidence that men with more than one older brother are more likely to be homosexual than their eldest brother to argue that homosexuality in males can be caused by exposure of the fetus to higher levels of androgen, not lower levels, as Dörner and Money had hypothesized. Thus Williams and colleagues (2000) reasoned that elevated androgen levels prenatally influence both sexes to become homosexual. One problem with this argument is that the male fetus itself produces much more testosterone than any amounts contributed to it by its mother. The other problem is that birth order determines a range of influences on development, especially after birth when older and younger children have differing roles in the family. The paper made no attempt to discuss cultural aspects of being in a family and gay.

One is reminded here of the many decades of earnest research activities from the nineteenth century to the present, measuring differences in brain sizes, in sizes of specific nuclei in the brain, and in anatomical and genital size differences for the sake of identifying weaknesses, inferiority, or abnormality in women, people of color, or homosexuals. These ignoble efforts have been doomed to failure. They have not shed light either on sex differences or on homosexuality. In the case of the finger length study, the leap from finger size to hormones and to sexual orientation is a jump into the world of the fanciful.

As we discussed earlier, previous research had shown no greater incidence of lesbian behavior in women exposed to exceptionally high levels of androgens (i.e., with adrenal hyperplasia syndrome) during their fetal development. A palmist might have interpreted the finger length as a sign of nonconventionality or of artistic leanings, or might have concluded that the different length of fingers was a result of permanent stress. Those correlations could be spurious or real. We will never know because it would require testing a large random selection of human beings in different regions of the world for the same traits, and then, when we have the results, where would we be? We were puzzled to find that the results were reported as an item of great interest to science. The absence of any clear statement of the methods used and control for orientation of the fingers makes it even more remarkable that the paper was published in *Nature*, a scientific journal usually thought of as reserved for reports on the most outstanding scientific work available. In our opinion, one cannot seriously entertain the possibility that such results get wide coverage for their scientific merit. Grasping at straws to establish genetic causes reflects a social climate that seeks certainty, or, more unpleasantly, might want to find the mark of Cain written clearly on someone's forehead as a way of detecting the rule breaker in its midst.

Other studies (Sanders and Kadam 2001, Sanders and Waters 2001) have related the number of ridges in fingerprints to sex differences in performance of cognitive tasks and then to assumed differences in androgen exposure of the fetus. The number of dermal ridges on the fingertips is asymmetrical, such that most individuals have more on the fingers of the right hand than on the left hand. These studies show that having more ridges on the fingers of the right hand is associated with a male pattern of cognition, whereas having more ridges on the left fingers is associated with a female cognitive pattern, irrespective of sex. But—and here is the catch—more ridges on the left fingers is more common in females. Also males have more ridges in total. The cause of these differences has been assumed to be exposure to sex hormones during early development (Sanders and Waters 2001), or genes, or both (Green and Young 2000). Not surprisingly, comparisons have been made between the patterns in heterosexuals and homosexuals as well as in transsexuals. Green and Young (2000) found that the finger ridge asymmetries differed in homosexuals versus heterosexuals but they were not related to transsexualism.

Have researchers found convenient measures for sex differences and sexual orientation in relative finger length and ridge patterns? Perhaps, but we note the marginality of the differences and the large overlaps. It is more important to look at the social implications of these results if they are to be taken seriously. Whether or not the data are correct, they lay the basis for individuals to be labeled as homosexual from a mere fingerprint or handprint. False accusations are likely, or, if the method of distinguishing between heterosexual and homosexual could ever be established, homosexuals may be forced to "come out" against their will. Added to this, finger and hand distinguishing marks are evident at birth and even before it. What will that mean for the owners of hands and fingers that do not conform to the perceived heterosexual pattern? Society or parents not accepting of homosexuality may opt for aborting the fetus seen to be so afflicted or for raising the child in ways so as to prevent the expression of homosexuality, no matter how oppressive these might be. The hunt is on for gay labels in genes, fingers, or hands. We might as well ask why finding these labels is of such interest. It would not be far-fetched to suspect that it is not mere scientific curiosity. (For further discussion, see Birke 2002.)

One does not have to be on the side of either a "rhetoric of concern" or a "rhetoric of promise" to see that the gay gene issue has more than scientific interest stamped all over it (Conrad and Markens 2001), and in this constellation one needs to scrutinize the science even more carefully than one ought to in every other case. It is a mark of a good, honest scientist to be clear about the goals of a piece of research before commencing any study and to be willing to admit the limitations of the methods and results.

Even though animals of many species engage in homosexual practice, there is no gene for homosexuality. Sexual events of this kind, now documented for species as different as leopard seals and pygmy chimpanzees, neither prevent reproductive matings nor are commonly associated with cues of reproductive status. Indeed, such sexual activity usually lies outside estrus and reproductive seasons. They are typically associated with very strong social bonds but cannot easily be associated with "selfishness" of genes. There is so far only one known example of male to male matings being switched on by one gene. This is the fruit fly (drosophila) and, ironically, much interest was expressed in its male–male mating behavior at the very time when the "gay gene" thesis was proposed and hotly debated. The fruit fly is an invertebrate species and its mechanisms are simpler to study than those of a vertebrate species. One would be hard pressed to call a fruit fly homosexual, but it can show behavior of persistent male–male interest under certain circumstances. One notes that it is not the action of a gene alone that dictates the male–male mounting, regardless of circumstances. On the contrary: the behavior is dependent on specific environmental conditions, such as a rise in temperature, that elicits a suite of biochemical reactions with the result of a shift in attention away from females and toward other males (Kitamoto 2002, Zhang and Odenwald 1995). Although interesting, it is an entirely different matter to assume that finding this gene might lead us to discovery of the gene (g) for homosexuality in humans. It is far from likely that the same gene is present in vertebrates to have an effect on sexual behavior, let alone in humans (Thompson 1995).

Despite the availability of new technologies in molecular genetics, we have no new and convincing evidence that links human behavior to the direct expression of the genes alone, as often stated by the general media and some scientists. These voices are the echoes of a society at pains to understand itself, and to do so in the most rigid and unflinching terms. They are an effort to recruit science into the social debate and to use it for purposes of control, not emancipation. Regrettably, the search for certainty of knowing ourselves has taken us to the most basic building blocks of our makeup. Certain subjects in the biological and psychological sciences are, as the biologist Ruth Hubbard (1994) has said, "suffused with cultural meanings and embedded in power relationships."

10

Complexity, Plasticity

In *Professions for Women* (1921), Virginia Woolf expressed dismay with her guardian angel. The guardian angel in her house was a defender of the status quo, a conservative soul living by a specific code "in the best interest" of her charge. When Virginia Woolf tried to write she encountered her guardian angel with the very first words: "The shadow of her wings fell on my page; I heard the rustling of her skirts in the room," and the reader thinks that the angel will make her writer famous. The angel says: "My dear, you are a young woman. You are writing about a book that has been written by a man. Be sympathetic, be tender, flatter, deceive, use all the wiles of our sex. Never let anyone guess you have a mind of your own." And at those words Woolf killed her own angel, the ultimate act of blasphemy.

The "whisperings within" (Barash 1979) were not uttered by the genes, of knowing her "true nature"; it was a knowledge of the sanctions and rules of her social world which, under ordinary circumstances, would prevent her from writing what she actually thought. As Marcuse has said, and we mentioned before, liberation from existing social process lies in the dynamic of negativity, that is, in the negation of what exists, and in making the absent present because the "greater part of the truth is in that which is absent" (cf. Cornell and Thurschwell 1987; Marcuse 1968).

The power within to defy convention is a power of human spirit. It is the act of self-creation, the freedom of will not to do what is expected, not to be "clever" by exploiting what is but by pushing the boundaries beyond what seems immutable. And as the boundaries shift and change so does the spirit of resistance, now aware of false guardian angels and of the risks of claiming essentialist identities. Much later, Donna Haraway, along with her generation, had to create her own "blasphemy" by declaring herself a monstrous creation that challenges all binary oppositions (first world–third world; male–female; science–art; domination–subjugation), and every desire for wholeness (cf. Sandoval 2000). She, too, if in very different garb and imagery, lives both a social reality and a fiction. "Wholeness," in these new cyborg and anticolonialist feminisms, is a dangerous business because the only wholeness that is available is in the garb of traditional forms, precincts, and images, and women cannot think and act to free themselves unless they step out of those images and forms and away from narratives and realities of power.

From the essay by Virginia Woolf to Donna Haraway's "A Manifesto for Cyborgs" (1985; see also Haraway 1982, 1991) and to this day lies the greatest collective intellectual and cultural oeuvre ever created by women in the entire recorded history of humanity. This has occurred in spite of scientific opinion on women over two hundred years and philosophical traditions much older. It begs the question of what we ought to think of theoretical positions that argue that women are either biologically not capable of acting or not designed to do as well as men and over the same range of endeavors. We have been at pains to argue that it is not a matter of adopting a middle ground between nature and nurture. Instead, both evolution and development are processes of becoming; this occurs within each individual in a dynamic and dialectic process.

Those who wish to argue that we are but a product of our genes and nothing more (even if other factors are acknowledged) have declared a bankruptcy of the human spirit, a denial of libido even, a denial of consciousness, the subconscious, and one of the most important and difficult aspects of human life altogether: the power to choose. Humans have a passion for inventing things that lift us beyond our human limitations. We build airplanes to overcome flightlessness. We fly to the moon even though we need to breathe air and rely on gravity. We have invented submarines to take us under water. In every way human inventiveness has taken us beyond our own limits and powers, at least in the physical realm. Yet when it comes to perceptions of sexual differences or sexual orientation, generation after generation of scientists (usually men) have attempted to prove the very things that women are trying to overcome: our limitations. They say that women were meant to be reconciled to being confined to their childbearing role, and homosexuality has to have a biological cause.

We do not have to accept "isms" of any kind, whether or not they are linked to biology in some way. Ageism, for instance, takes it for granted that old people cannot learn, cannot be creative, and can no longer think clearly. Countless examples show that the human brain retains a plasticity to learn well into old age, that some of the greatest artistic achievements emerged from old bodies (mature but not old minds). We do not have to accept racism because different colors of hair, eyes, and skin are merely a mark of our diversity as a species. Variability and diversity are nature's greatest assets. We do not have to accept sexism because the so-called proof for women's "inferiority" has thus far always been based on pseudoscience. It has been one of the longest and hardest-to-kill frauds and one of the lowest points of human insight.

The malleability of humans is borderless, and that is perhaps one of our greatest gifts. In biology, this would be called adaptability. We do not have to accept the mythology of hunter–gatherer minds. As far back as 1967 Gottfried Kurth argued that human development over a two-million-year span required plasticity and a reduction of "instincts." The concept of instinct has been given a range of different meanings (see Bateson 2000) but all imply behavior without learning. Yet there is very little that we do without first having to learn, and in every aspect of its development the brain relies on experience. We have not been adapted in the way evolutionary psychologists wish to see it, but are much more varied and capable of responding to new challenges. We are not ancient mammals in a brave New World, as evolutionary psychologists Palmer and Palmer (2002) wish to see us, trying to manage in a world for which we are not adapted. The human species may be one of the least specialized species on the planet, and malleability may have been its main survival. Moreover, malleability is a quality also present in many animals. Animals as well as humans need to learn from parents and from experience.

Sociobiology has moved on, and in the early twenty-first century has begun to include "epigenetic" phenomena that, in the old language, translated into environmental factors. Evolutionary psychology has generally not taken this step and has repeated many of the heresies of sociobiology without apology or remorse. It is thriving in some of the most elitist intellectual institutions and, together with molecular genetics, has had monetary support of a magnitude that can only be described as extraordinary (including from governments). Evolutionary psychology is a sign of the 1990s and the early twenty-first century. Its orthodox position has argued that the human mind contains discrete modules with separate functions under genetic control, although the genetics here can sometimes function as a necessary adjunct for given traits and, in other cases, as a totally self-sufficient explanation for a trait. Thus far, the idea of modules is only a hypothesis on how the brain may process information and one particularly attractive in conjunction with mathematical and computer metaphors in the midst of a computer age and artificial intelligence. A paper that we cited in the fourth chapter begins by saying that "the human brain is a set of computational machines, each of which was designed by natural selection to solve adaptive problems faced by our hunter–gatherer ancestors" and that "these machines are adaptive specializations: systems equipped with design features that are organized such that they solve an ancestral problem reliably, economically and efficiently" (Duchaine et al. 2001, p. 225). If the imagery remained within computer language, one might even concede that the comparison may have some merit, but it is hard to find a reasonable explanation, let alone a scientific one, why stone-age people invade the computer screen. Thus, when evolutionary psychology equates human activity with biological preparedness, and then goes further to argue that manhood and womanhood are genetic dispositions that have been prefigured according to an ancestral blueprint (cf. Tooby and Cosmides 1990), then we are facing some serious problems of ideology.

The excesses of Rushton's (1994, 1999) openly and defiantly racist agenda notwithstanding (as we perceive it, although he would probably not regard his ideas as racist but as true), evolutionary psychology has hardened its position into a supremacist standpoint. Evolutionary psychologists and their supporters do not lack publishers and eminent people for praise. For instance, a book by Deborah Blum (1997), *Sex on the Brain*, is praised by E. O. Wilson on the flap of the book as a "carefully researched, factually based" book that is "marked less by ideology and resentment and more by frank and joyful exchange." Despite the fact that the book is riddled with distortions and suffers from a worryingly cavalier approach to science and to research findings (see the book review by Melissa Hines in *Nature*, 1998), it is telling that her work fits so well into the new ideological push to the extreme right. Blum argues (unchallenged by many) that being blonde has a mating advantage because men prefer it. So men prefer blondes as mates (and this is, of course, not at all a manipulated view but has a biological basis)? Do they really? That would mean that over 90 percent of males (or females) would never get the partner of their choice because blond hair, unless out of a bottle, is naturally in short supply. How, other than for racial reasons, would anybody want to elevate paleness and blondness to the most desired attributes for sexual partnership? What is evident from the blond and blue-eyed imagery in advertising, in the media, and in Blum's book, is a push for white supremacy. And, apparently, we are meant to believe that this is not a matter of ideology. Ultimately, supremacist viewpoints provide justifications for actions in the broader social world but, by and large, they do not contribute to a progress of insight and science.

This logic, as we have tried to show, can lead to a conceptually totalitarian approach to all matters of life, using (not surprisingly) totalizing categories such as women and blacks (Lather 1991). In the final analysis, evolutionary psychology debates have had too much in common with biologisms expounded in the widely lauded and supported intellectual positions of eugenics, racism, and fascism of the 1920s and 1930s, even though some individual researchers in the field may not then and would not now believe this, nor would they act in ways that overtly support this. Throughout all ages scientists have claimed the right to carry out their work regardless of social consequences and even be innocent or unaware of such impact. Hannah Arendt said succinctly that some of the greatest evil has been done by people who think neither good nor evil (Arendt 1963, Kaplan and Kessler 1989). It took a world war, the death of 55 million people, a holocaust, and a reconstitution of the senses from total oblivion to make these theories of *Menschenunwert* (human trash) and racist arguments for biologically based "healthy gene pools" shrink back and return into some dark crevices of the earth. Now we are back again with theories of biologisms that seem to conjure up some of the old demons. Camus' novel *The Plague* comes to mind.

Worse, there is a streetcar, not named desire, that promises a ride into a world of ultimate answers and ultimate truths. Excitement is generated in the expectation that evolutionary psychology in concert with the new molecular genetics will finally crack the code of all codes and bring home the final prize of the science of life: the truth about all living things. In some quarters, evolutionary psychology is now hailed as a truly progressive force (Brase 2002).

The problem is that any serious scientist would eventually have to admit that matters of genes, hormones, environment, and their interactions are extremely complex and do not lend themselves to simple unidirectional theories of causation. Some of the scientists writing with such scorn and panache about sex differences, genetic causation of crime, homelessness, and antisocial behavior may also have to agree that their views are not value free and not free from ideology. Yet this is an impression hard to shake when a book review of Hamer and Copeland's popular version of the gay gene research, digresses into the following diatribe:

> Hamer will make no friends among the radical feminist/French theory/postmodern/queer studies/MLA/academic chic crowd, but then, such ideologues tolerate only devotees, not independently-minded colleagues. Hamer is nothing if not an independent and expansive thinker. [Gonsiorek 1995, p. 262]

After the shock of this statement, one notes that, as we detailed in Chapter 9, the gay gene claims were not substantiated and if scientists make "no friends" it is usually because their science is not rigorous enough.

In the power struggle of various scientific disciplines for eminence (Fausto-Sterling 2001) there is a major price tag on clinching answers to questions that have puzzled the human mind for thousands of years.

The first recorded philosophical writings on sex differences can be found in ancient Greece. Reading some of them makes one wonder how impressed Aristotle and Plato would be if we invited them to preside over a class on sex differences today. Perhaps gender schemas have not appreciably changed, even though we now have more research tools available. Hines (1998) said so nicely in her review of Blum's book:

> The trouble with sex is that people, scientists and non-scientists alike, have their own theories about how men and women differ. Developmental psychologists call these theories "gender schemas" and find that they influence memory. For example, children shown pictures of male nurses and female doctors later confidently recall seeing female nurses and male doctors. Those writing popular books on psychological sex differences often seem to have similar problems with remembering research findings. [p. 146]

There are many things we do not fully understand. Some of the answers to questions that scientists are seeking now may well be around the corner. But as long as scientific evidence cannot be procured, it is premature to make claims about the outcomes. And when the outcomes do not fit preexisting ideological positions, no attempt should be made to bend the findings to fit them. The opposite should happen: the theories should be reviewed to fit the findings. Interpretation of evidence is an important part of science. If some scientists are more skeptical than others and require more rigorous evidence, this too is part of science and the furtherance of knowledge. Skepticism and rigorous science are not bad faults compared to moving prematurely to conclusions, especially when they influence social attitudes.

The onward march of evolutionary psychology, not just as a separate discipline, but also as the *master* discipline that intends to eventually integrate and swallow up all others, is already in progress. It has borrowed from psychology and evolutionary biology, has stumbled blithely and somewhat blindly (unskilled in some cases) into anthropology. It has started to colonize sociology (Rose 2000) and it has further taken it upon itself to attempt to squelch anyone whose opinions might not agree with the positions of evolutionary psychology. The entire Western intellectual history and philosophy is reduced to the term "culturalism" (from Aristotle to Descartes, from Locke to Adorno, from Plato to Foucault), a self-positioning that suggests implicitly that no thought before evolutionary psychology and biology has had anything to contribute to any "real" questions (Peritore 1997). An example of a new mob culture of opinion makers—aggressive, often unscientific, and at times not observing the minimal codes of accepted academic politeness—is the response to a book by Rose and Rose, *Alas Poor Darwin* (2000), edited and partly written by two eminently qualified scholars, a team of scientists cum social scientists. We have followed some of these responses with concern (Kurzban 2002) because the manner in which the rebuttals have been conducted seems to us to show marks of irrationality, common to religious fundamentalism and to ideology. The personalized, nearly defamatory, attacks on the book and its writers should not be mistaken for a passion for science. They are, instead, a passion over a particular world-view and that, too, is not new.

Evolutionary psychologists may rightly criticize the social sciences for several failings in their interpretations of the world, as one may rightly criticize scientific endeavors for their biases. In this last chapter we wish to respond to some of these criticisms as far as they are relevant to genes, gender, and sex.

Peritore (1997) has written an evolutionary critique on "culturalism" and in it, he takes "culturalism" and social science to task. Published, appropriately, in the journal *Research in Biopolitics*, his essay argues through a number of issues that pertain directly to our exposition in this book and should therefore be answered. First, he charges the social sciences with being "metaphysical" and thus, presumably, divorced from a (biological) reality. He is correct in saying that social science has long been grounded on notions of the superiority of humans to animals, if one reads superiority as supremacy and uniqueness. No doubt, such attitudes have been predominant paradigms in many cultures, but we need to remind those critics pointing the finger at social science that such attitudes were not just the province of one cluster of disciplines. Indeed, the sciences of the nineteenth century, in particular, and in concert with imperialism, vigorously supported the idea of human supremacy qua white supremacy. Indeed, racism and sexism received legitimation primarily via science. We have presented evidence of scientific writing in this book exemplifying that science, chiefly biology, was intent on convincing the world that the white human male was the pinnacle of all creations. Women and colored people were seen to be at a lower evolutionary level and some "natives" were even regarded as being closer to apes, according to these "scientific" views. Pseudoscience created spurious evolutionary links and translated these into assumed "natural" hierarchies within the human species. Current science and culture are well aware of the continuities between animals and humans (Rogers and Kaplan 2000). We have discovered traits that appeared early in the evolutionary development, such as lateralization of the brain (Rogers and Andrew 2002). We know that a number of qualities, such as problem solving, tool use, spatial memory, and other abilities, once thought to be uniquely human, are traits which, perhaps rudimentarily, appear in some avian and primate species (Byrne 1995, Evans 1997, Rogers 1997). Knowing this today, within science, also means that the "not-knowing" before was also part of science and that science, like any other discipline, believed in the uniqueness of the human primate. The modern scientific agenda is to show when and how certain traits developed, how and why they persisted, and in which species. In the arena of science and sex differences, women have had to free themselves from the shackles of biologisms which, for alleged scientific reasons, relegated them into inferior positions within the "natural" scheme of things.

Peritore further argues that Western cultural theory has presupposed that policy can modify "human nature" and thereby freely create alternative futures, and that these premises run counter to evolutionary biology. These views are therefore allegedly untenable (presumably meaning that evolutionary biology holds the truth). There are two flaws in this argument. First, policy in and of itself has never been understood as changing "human nature." At worst, it is designed to have a punitive or controlling function and at best, it is conceived of as a vehicle to permit greater achievement and freedom of individuals. Educational language is full of metaphors referring to "fulfilling one's *potential*"; this clearly has nothing whatever to do with "modifying nature." That women are now ranked among the finest minds is not because women had to modify "their nature." Like Virginia Woolf, however, these women may have had to kill the angel of convention to be what they wanted to be, instead of what convention told them to be. (See, for example, Keller's 1983 account of the life and work of Nobel Laureate Barbara McClintock.)

Second, implied in Peritore's argument is that evolutionary biology holds the key to life, while culture is in a discordant opposition to biological facts. This is an entirely unfounded assumption, at least with respect to gender. Cultures have persistently based their values and norms on reproductive sex differences and assigned roles to males and females narrowly defined by biology. That many of the limitations imposed on women and justified on the basis of their biology was proved to be well beyond biological dictates was conveniently overlooked, and the extensive edifice that has surrounded the construction of biology was meant to confirm, not deny, biology. Only in contemporary society has a full challenge been mounted to expose how much of the social rules that governed gender can actually be attributed to sex differences.

Third, "culturalism" has apparently saddled the horse in the wrong direction. Peritore (1997) writes that "culturalism violates scientific method in reversing the order of causality. In all organisms, without exception, information flows from DNA to protein to phenotype, and this flow is not reversible" (p. 115).

Here we are probably at the crux of the argument that has created insurmountable barriers between the various scientific and cultural perspectives. We have outlined throughout this book that specific experiential factors have a bearing on the individual organism. Rogers (1982) found, for instance, that exposure to light of one eye of an avian embryo can permanently reverse some expressions of brain lateralization. We described in chapter 3 how different treatments of male and female rat pups altered their sexual behavior as adults. Experience cannot alter the DNA itself but it can alter what genes are expressed and the biology of the brain (its structure and cellular functioning), as well as the actual secretion of hormones. This is not a unidirectional process but one that is dialectic. Moreover, interactions between developmentally specified versus developmentally modifiable events occur at *every stage of development*. The possible number of modifications may be circumscribed to some extent, and here genes are involved, but it is still a process that is not unidirectional. Genes have a role in determining the secretion of sex hormones, although this role is not absolute. There are, of course, many other ways in which genes exert their influence, but always in interaction with other biological factors and experience. At no stage in the process of development would it be possible to single out an event and say that it is entirely caused by genes, or entirely caused by sex hormones, or entirely the result of experience. Nor is it possible to see the effects of genes and experience as additive in any simple way. The models examined here show that these various contributions to development are inseparable at all stages. We have also pointed out that culture (at the symbolic level) has a way of exerting its influence on the individual. We are not just referring to a single loop between genes and environment at the stimulus–response level, but to a second loop from symbolic culture to the individual and vice versa. In fact, even these two loops are blurred by interactions. To draw the conclusion, as Peritore seems to think one must, that any researcher who places some emphasis on learning and development a priori believes that environmental influences predominate over genes, is actually incorrect. One does not have to believe this at all.

Finally, the assumption of unidirectionality harbors one of the most consistent flaws in genetic explanations of social life. Directionality (allegedly one-way from genes to society) is mistaken as *causality*. This is a serious flaw indeed. As others have argued before (Elredge and Tattersall 1982, Mandler 1997), logically there is no way in which narrowly distributed properties can be explained in terms of the more broadly distributed ones. For instance, the fact that all living things consist largely of water (widely distributed properties) has no explanatory power in why most southern Europeans are Catholic (narrowly distributed properties). Logically, water could be replaced by genes and we have the same inplausibility problem. We may share 40 percent of our DNA with a banana, but no one would expect us to wish to bond with bananas on the grounds of some shared properties.

On a final note, we want to end this book by laying to rest one of the most reductionist of all assumptions: that the sum total of meaning of human life is explicable in terms of reproductive success and that culture and other human activities are background variables. If this were so, the entire human endeavor and its forays into culture and even technology would have to be one of the worst evolutionary mistakes. Indeed, it would have to be a maladaptation to the demands of reproductive success because culture diverts enormous resources to activities that are not reproductively essential. We might even say, with Freud, that culture is a force that represses and channels the sexual drive and therefore limits and diverts these energies into other activities not associated with reproduction. Indeed, as a species, we have increasingly moved away from the centrality of reproduction in a life cycle. We live longer, and therefore procreation takes up a shorter part of the life cycle. We are healthier, so more newborns stay alive. Humans in privileged nations do not have to reproduce continually.

If there is one single deep human need, it is freedom and, if there is one road that we have traveled as a species, it is to find ways to give greater expression to that freedom and to be safe from the traumatic vicissitudes of life. Freedom is as much *from* many things (as fear of being killed, of starvation) as it is *to* many things. The explorations of freedoms *to* many things are still in the social laboratory; there are some women who have considerably more freedoms than others. Neo-Darwinians would believe that all human society across the world and at all times has ultimately organized itself around the core activity of reproductive success, and any alternative social organizations were ultimately short-lived and at very great cost (Peritore 1997). However, there is much more enduring evidence that suppression of freedom at national and individual levels has always been the most costly and bloody business in human history. The killing of a foe for the sake of taking his/her freedom and life itself is not clever from an evolutionary point of view but it is clever for maintaining power. There are so many human activities that do not support the notion that reproduction is the core and genetically driven activity of our species. The only times when reproduction has been considered an explicit political and cultural goal were at times of war and at times of strong racist views.

The question is whose interests do neo-Darwinian theories serve. Here we might refer back to Habermas (1971), who introduced the concept of "emancipatory interests" in his explorations of critical theory. He argued that all new knowledge has objects or goals to which it is applied, and he pointed out that the goal of prediction, in social science, has usually resulted in the production of knowledge for purposes of social control. Social sciences guided by emancipatory interests, however, are concerned with generating knowledge in order to liberate people from social domination.

The power to suppress freedom of any kind, even the freedom of love and laughter (as in Spanidou's novel *God's Snake*; see chapter 5) serves no purpose other than power (and not reproduction). Many needs, it seems, may not serve reproduction. Psychology and psychiatry were disciplines originally established in opposition to biology. Freud's famous letter to Carl Müller-Braunschweig emphasized strongly that a distinction had to be made between what is "psychic" and what is biological because human needs and hopes were not necessarily born of biological needs, including even love. Human yearnings have also grown beyond biology.

References

Aboitiz, F., Scheibel, A. B., and Zaidel, E. (1992). Morphometry of the Sylvian fissure and the corpus callosum, with emphasis on sex differences. *Brain* 115: 1521–1541.

Adams, C. E. (1988). *Women Clerks in Wilhelmine Germany: Issues of Class and Gender.* Cambridge, MA: Cambridge University Press.

Adorno, T. W. (1975). *Negative Dialektik.* Frankfurt/Main: Suhrkamp.

Adorno, T. W., and Horkheimer, M. (1979). *Dialectic of Enlightenment.* London: Verso.

Alcock, J. (2001). *The Triumph of Sociobiology.* Oxford: Oxford University Press.

Allen, L. S., Richey, M. F., Chai, Y. M., and Gorski, R. A. (1991). Sex differences in the corpus callosum of the living human being. *Journal of Neuroscience* 11: 933–942.

Allen, M. P. (1984). Fantasia Fair. *Australian Seahorse Bulletin.* June, p. 2.

Allen, S., and Gorski, R. A. (1992). Sexual orientation and the size of the anterior commissure in the human brain. *Proceedings of the National Academy of the Sciences USA* 89: 7199–7202.

Alley, R. (1957). *Peking Opera*. Peking: New World Press.

American Psychiatric Association (1994). *Diagnostic and Statistical Manual of Mental Disorders*. 4th ed. Washington, DC: Author.

Anastasi, A. (1966). *Differential Psychology: Individual and Group Differences in Behavior*. New York: Macmillan.

Anderson, B. (1983). *Imagined Communities: Reflections on the Origin and Spread of Nationalism*. London: Versa.

Andreae, S. (1998). *Anatomy of Desire: The Science and Psychology of Sex, Love and Marriage*. London: Little, Brown.

Apter, T. (1985). *Why Women Don't Have Wives: Professional Success and Motherhood*. Basingstoke, UK: Macmillan.

Archer, L. J., ed. (1988). *Slavery and Other Forms of Unfree Labour*. London: Routledge.

Arendt, H. (1958). *The Human Condition*. New York: Routledge, 1992.

——— (1963). *Eichmann in Jerusalem: A Report on the Banality of Evil*. New York: Viking.

——— (1972). *Crises of the Republic*. New York: Harcourt Brace Jovanovich.

Auger, A. P., Tetel, M. J., and McCarthy, M. M. (2000). Steroid receptor coactivator –1(SRC-1) mediates the development of sex-specific brain morphology and behavior. *Proceedings of the National Academy of Sciences* 97: 7551–7555.

Austin, C. R., and Short, R. V. (1984). *Reproductive Fitness: Reproduction in Mammals* 4, 2nd ed. Cambridge, UK: Cambridge University Press.

Bacchi, C. L. (1991). Pregnancy, the law and the meaning of equality. In *Equality, Politics and Gender*, ed. E. Meehan and S. Sevenhuijsen, pp. 71–87. London: Sage.

Bagemihl, B. (1999). *Biological Exuberance: Animal Homosexuality and Natural Diversity*. London: Profile Books.

Bailey, J. M., and Pillard, P. C. (1991). A genetic study of male sexual orientation. *Archives of General Psychiatry* 48: 1089–1096.

Bailey, J. M., Pillard, P. C., Neale, M. C., and Agyei, Y. (1993). Heritable factors influence sexual orientation in women. *Archives of General Psychiatry* 50: 217–223.

Baker, R. R. (1996). *Sperm Wars: Infidelity, Sexual Conflict and Other Bedroom Battles*. London: Fourth Estate.

Baker, R. R., and Bellis, M. A. (1989). Number of sperm in human ejaculate varies in accordance with sperm competition. *Animal Behaviour* 37: 867–869.

——— (1993). Human sperm competition: ejaculate manipulation by females and a function for female orgasm. *Animal Behaviour* 46: 887–909.

Baker, S. W., and Ehrhardt, A. A. (1974). Pre-natal androgen, intelligence, and cognitive sex differences. In *Sex Differences in Behavior*, ed. R. C. Friedman, R. M. Richart, and R. L. Vandeeds, pp. 53–84. New York: Wiley.

Ball, C. A. (2001). Essentialism and universalism in gay rights philosophy: liberalism meets queer theory. *Law and Social Inquiry* 26: 271–293.

Barash, D. (1979). *Sociobiology: The Whisperings Within*. New York: Harper & Row.

Barber, E. W. (1994). *Women's Work: The First 20,000 Years*. New York and London: Norton.

Bardwick, J. M. (1971). *Psychology of Women*. New York: Harper & Row.

Barkan, E. (1992). *The Retreat of Scientific Racism*. Cambridge, UK: Cambridge University Press.

Barkow, J., Cosmides, L., and Tooby, L., eds. (1992). *The Adapted Mind: Evolutionary Psychology and the Generation of Culture*. New York: Oxford University Press.

Barnard, A. (1999). Modern hunter-gatherers and early symbolic culture. In *The Evolution of Culture*, ed. R. Dunbar, C. Knight, and C. Power, pp. 50–70. New Brunswick, NJ: Rutgers University Press.

Barnes, M. (1988). Mathematics: a barrier for women? In *Crossing Boundaries: Feminisms and the Critique of Knowledges*, ed. B. Caine, E. A. Craz, and M. de Lepervanche, pp. 28–42. Sydney: Allen and Unwin.

Barr, J., and Birke, L. (1998). *Common Science? Women, Science and Knowledge*. Bloomington, IN: Indiana University Press.

Barron, R., and Norris, G. (1976). Sexual division and the dual labour market. In *Dependence and Exploitation in Work and Marriage*, ed. D. Barker and S. Allen, pp. 47–69. London: Longman.

Bateson, P. (2000). Taking the stink out of instinct. In *Alas, Poor Dar-*

win: Arguments Against Evolutionary Psychology, ed. H. Rose and S. Rose, pp. 189–207. New York: Harmony Books.

Baum, M. J., Bressler, S. C., Daum, M. C., et al. (1996). Ferret mothers provide more anogenital licking to male offspring: possible contribution to psychosexual differentiation. *Physiology and Behavior* 60:353–359.

Bäumer, G. (1904–1905). Was bedeutet in der deutschen Frauenbewegung "jüngere" and "ältere" Richtung? *Frau* 12: 321–328.

Baumgardner, J., and Richards, A. (2000). *Manifesta: Young Women, Feminism and the Future*. New York: Farrar, Straus and Giroux.

Beach, F. A. (1971). Hormonal factors controlling the differentiation, development and display of copulatory behavior in the ramstergig and related species. In *Biopsychology of Development*, ed. E. Tobach, L. R. Aronson, and E. Shaw, pp. 249–296. New York: Academic Press.

Beechey, V., and Perkins, T. (1987). *A Matter of Hours: Women, Part-Time Work and the Labour Market*. Cambridge, UK: Polity.

Bell, A. D., and Variend, S. (1985). Failure to demonstrate sexual dimorphism of the corpus callosum in childhood. *Journal of Anatomy* 143: 143–147.

Bell, R. (1985). *Holy Anorexia*. Chicago and London: University of Chicago Press.

Bem, S. L. (1974). The measurement of psychological androgyny. *Journal of Consulting and Clinical Psychology* 42: 155–162.

Benhabib, S. (1989). On contemporary feminist theory. *Dissent* Summer: 366–370.

Benjamin, J. (1995). Sameness and difference: "overinclusive" model gender development. *Psychoanalytical Inquiry* 15: 125–142.

Benjamin, W. (1970). *Illuminations*. London: Cape.

Berch, B. (1982). *The Endless Day: The Political Economy of Women and Work*. New York: Harcourt Brace Jovanovich.

Bereczkei, T., Voros, S., Gal, A., and Bernath, L. (1997). Resources, attractiveness, family commitment—reproductive decisions in human mate choice. *Ethology* 103: 681–699.

Berenbaum, S., and Hines, M. (1992). Early androgens are related to childhood sex-typed toy preference. *Psychological Science* 3: 203–206.

Berman, E. (1970). Cited by C. Lydon. Role of women sparks debate by Congress woman and doctors. *New York Times*, July 26, p. 35.

Berry, J. W. (1966). Temne and Eskimo perceptual skills. *International Journal of Psychology* 1: 207–229.

——— (1971). Ecological and social factors in spatial perceptual development. *Canadian Journal of Behavioural Science* 3: 324–336.

Bimonte, H. A., Fitch, R. H., and Denenberg, V. H. (2000). Neonatal estrogen blockade prevents normal callosal responsiveness to estradiol in adulthood. *Developmental Brain Research* 122: 149–155.

Birke, L. (1979). Is homosexuality hormonally determined? *Journal of Homosexuality* 6: 35–50.

——— (1982). Cleaving the mind. In *Against Biological Determinism*, ed. S. Rose, pp. 60–78. London: Alison Busby.

——— (2002). Unusual fingers: scientific study of sexual orientation. In *Handbook of Lesbian and Gay Studies*, ed. D. Richardson and S. Seidman, pp. 55–72. London: Sage.

Birrell, R. (1995). Women storm into the top professions. *The Australian* 27: 1.

Bishop, K. M., and Wahlsten, D. (1997). Sex differences in the human corpus callosum: myth or reality? *Neuroscience and Biobehavioral Reviews* 21: 581–601.

Bittman, M. (1991). *Juggling Time: How Australian Families Use Time.* Canberra: Office of the Status of Women, Department of the Prime Minister and Cabinet.

Blaustein, J. D., and Osler, D. H. (1989). Gonadal steroid hormone receptors and social behaviours. In *Advances in Comparative and Environmental Physiology Vol. 3: Molecular and Cellular Basis of Social Behavior in Vertebrates*, ed. J. Balthazart, pp. 31–104. Berlin: Springer-Verlag.

Bleier, R. (1984). *Science and Gender*. New York: Pergamon.

Bloch, E. (1985). *Werkausgabe (Collected Works)*. Frankfurt/Main: Suhrkamp Taschenbuch.

Bloom, H. (2000). *Global Brain: The Evolution of Mass Mind: From the Big Bang to the 21st Century*. New York: Wiley.

——— (2001). Instant evolution: the influence of the city on human genes: a speculative case. *New Ideas in Psychology* 19: 203–220.

Bluestone, N. H. (1987). *Women and the Ideal Society: Plato's Republic and Modern Myths of Gender*. Amherst: University of Massachussetts Press.

Blum, D. (1997). *Sex on the Brain: The Biological Differences between Men and Women*. New York and London: Viking Penguin.

Bock, G. (1984). Racism and sexism in Nazi Germany: motherhood, compulsory sterilization, and the state. In *When Biology Became Destiny: Women in Weimar and Nazi Germany*, ed. R. Bridenthal, A. Grossmann, and M. Kaplan, pp. 400–421. New York: Monthly Review.

——— (1986). *Zwangssterilisation im Nationalsozialismus* (Forced Sterilization under National Socialism). Opladen: Westdeutscher Verlag.

Bono, P., and Kemp, S., eds. (1991). *Italian Feminist Thought: A Reader*. London: Basil Blackwell.

Bordo, S. (1988). Anorexia nervosa: psychopathology as the crystallization of culture. In *Feminism and Foucault: Reflections on Resistance*, ed. I. Diamond and L. Quinby, pp. 87–118. Boston: Northeastern University Press.

Bourdieu, P. (1977). *Outline of a Theory of Practice*. London: Cambridge University Press.

Bower, B. (1993). Genetic clue to male homosexuality emerges. *Science News* 144(3): 37.

——— (1999). DNA's evolutionary dilemma. *Science News* 155 (6): 89.

Bowles, S., and Gintis, H. (1976). *Schooling in Capitalist America: Education Reform and the Contradictions of Economic Life*. London: Routledge & Kegan Paul.

Bradbury, J. W., and Vehrencamp, S. L. (1998). *Principles of Animal Communication*. Sutherland, MA: Sinauer.

Bradshaw, J. L. (1989). *Hemispheric Specialization and Psychological Function*. Chichester, UK: Wiley.

Bradshaw, J. L., and Mattingley, J. B. (1995). *Clinical Neuropsychology: Behavioral and Brain Science*. San Diego: Academic Press.

Bradshaw, J. L., and Rogers, L. J. (1993). *The Evolution of Lateral Asymmetries, Language, Tool Use, and Intellect*. San Diego: Academic Press.

Braidotti, R. (1993). Origin and development of gender studies in Western Europe. In *Establishing Gender Studies in Central and East-*

ern European Countries, pp. 23–32. Workshop in Wassenaar, The Netherlands (1993): Council of Europe. Woman's Network for Women's Studies, Zoetermeer.

Brase, G. (2002). Conceptual challenges (Book review of *Evolutionary Psychology: Innovative Research Strategies*, ed. Harmon R. Holcomb III). *The Human Nature Review* 2: 147–152.

Breedlove, S. M. (1992). Sexual differentiation of the brain and behavior. In *Behavioral Endocrinology*, ed. J. B. Becker, S. M. Breedlove, and D. Crews, pp. 39–68. Cambridge, MA: MIT Press.

——— (1997). Sex on the brain. *Nature* 389: 801.

——— (1999). The orthodox view of brain sexual differentiation. *Brain, Behavior and Evolution* 54: 8–14.

Breines, W., Carullo, M., and Stacey, J. (1978). Social biology, family studies and the antifeminist backlash. *Feminist Studies* 4: 43–67.

Bridenthal, R., Grossmann, A., and Kaplan, M., eds. (1984). *When Biology Became Destiny: Women in Weimar and Nazi Germany*. New York: Monthly Review.

Broca, P. (1873). Sur les crânes de la caverne de l'Homme-Mort (Lozère). *Revue d'Anthropologie* 2: 1–53.

Brown, A., and Finkelhor, D. (1986). Impact of child sexual abuse: a review of the research. *Psychological Bulletin* 99: 66–77.

Brown, G. R. (2001). Sex-biased investment in nonhuman primates: Can Trivers and Willard's theory be tested? *Animal Behaviour* 61: 683–694.

Brown, G. R., and Dixson, A. F. (2000). The development of behavioural sex differences in infant rhesus macaques (*Macaca mulatta*). *Primates* 41: 63–77.

Brown, T. J., Naftolin, F., and MacLusky, N. J. (1992). Sex differences in estrogen receptor binding in the rat hypothalamus: effects of subsaturating pulses of estradiol. *Brain Research* 578: 129–134.

Buffery, A. W. H. (1981). Male and female brain structure. In *Australian Women: Feminist Perspectives*, ed. N. Grieve and P. Grimshaw, pp. 58–156. Melbourne: Oxford University Press.

Buffery, A. W. H., and Gray, J. A. (1972). Sex differences in the development of spatial and linguistic skills. In *Gender Differences: Their Ontogeny and Significance*, ed. C. Ounsted and D. C. Taylor, pp. 123–158. Edinburgh: Churchill-Livingstone.

Burstyn, V. (1983). Masculine dominance and the state. In *Socialist*

Register, ed. R. E. Milliband and J. Santle, pp. 45–89. London: Merlin.

Burton, C. (1991). *The Promise and the Price: The Struggle for Equal Opportunity in Women's Employment.* Sydney: Allen and Unwin.

Buss, D. (1994). *The Evolution of Desire: Strategies for Human Mating.* New York: Basic Books.

——— (1999). *Evolutionary Psychology: The New Science of the Mind.* Needham Heights, MA: Allyn & Bacon.

Butler, J. (1990). *Gender Trouble.* New York: Routledge.

Byne, W., and Parsons, B. (1993). Human sexual orientation. *Archives of General Psychiatry* 50: 228–239.

Byrne, R. (1995). *The Thinking Ape: The Evolutionary Origins of Intelligence.* Oxford, UK: Oxford University Press.

Campbell, A. (2002). *A Mind of Her Own: The Evolutionary Psychology of Women.* Oxford, UK: Oxford University Press.

Campenni, C. E. (1999). Gender stereotyping of children's toys: a comparison of parents and nonparents. *Sex Roles* 40: 121–138.

Cantor, D. (1993). Items of property. In *The Great Ape Project*, ed. P. Cavalieri and P. Singer, pp. 280–290. London: Fourth Estate.

Caplan, P., MacPherson, G., and Tobin, P. (1985). Do sex-related differences in spatial abilities exist? *American Psychologist* 40: 786–799.

Carr, A. (1993). Loving our gay genes. M.S.O. July 28, p. 4.

Carruthers, F. (1997). Women lose battle of the boardrooms. *The Weekend Australian* 1: 6.

Carter, C. S. (1992). Hormonal influences on human sexual behavior. In *Behavioral Endocrinology*, ed. J. B. Becker, S. M. Breedlove, and D. Crews, pp. 131–142. Cambridge, MA: MIT Press.

Cartwright, J. (2001). *Evolution and Human Behavior.* Suffolk, UK: Bradford Books.

Casey, G. J. (1991). Eleanor Leacock, Marion Harris, and the struggle over warfare in anthropology. In *On Peace, War, and Gender: A Challenge to Genetic Explanations*, ed. A. E. Hunter, series eds. B. Rossof and E. Tobach, pp. 1–33. New York: The Feminist Press.

Casey, M. B., Nuttall, R., Pezaris, E., and Benbow, C. P. (1995). The influence of spatial ability on gender differences in mathematics college entrance test scores across diverse examples. *Developmental Psychology* 31: 697–705.

Cattell, R. B. (1987). *Beyondism: Religion from Science*, 2nd ed. New York: Praeger.
Chesler, P. (1972). *Women and Madness*. New York: Avon.
Chipman, S. F., and Wilson, D. M. (1985). Understanding mathematics course enrollment and mathematics achievement: a synthesis of the research. In *Women and Mathematics: Balancing the Equation*, ed. S. F. Chipman, L. R. Brush, and D. M. Wilson, pp. 275–328. Hillsdale, NJ: Erlbaum.
Choleris, E., and Kavaliers, M. (1999). Social learning in animals: sex differences and neurobiological analysis. *Pharmacology, Biochemistry and Behavior* 64: 767–776.
Chorover, S. L. (1979). *From Genesis to Genocide. The Meaning of Human Nature and the Power of Behavior Control.* Cambridge, MA: MIT Press.
Churchill, W. (1967). *Homosexual Behaviour among Males: A Cross-Cultural and Cross-Species Investigation.* Englewood Cliffs, NJ: Prentice Hall.
Cicero, M. T. (1951). *De Legibus*, vol. I, trans. C. W. Keyes. London: Heinemann.
Cixous, H. (1975a). Le rire de la Meduse. *L'Arc* 28: 39–56.
——— (1975b). *La Jeune née*. Paris: U. G. E.
——— (1986). *Entre l'écriture*. Paris: des femmes.
——— (1987). *Le Livre de Promethea.* Paris: Gallimard.
Clarke, E. H. (1873). *Sex in Education: or A Fair Chance for Girls*. Boston: J. R. Osgood.
Clarke, L. (1979). Women and Locke: Who owns the apples in the Garden of Eden? In *The Sexism of Social and Political Theory: Women and Reproduction from Plato to Nietzsche*, ed. L. Clarke and L. Lange, pp. 16–40. Toronto: University of Toronto Press.
Cleary, N., and Von Mueffling, D. (2001). *The Art and Power of Being a Lady*. New York: Atlantic Monthly.
Clutterback, D., and Devine, M., eds. (1987). *Business Women: Present and Future*. London: Macmillan.
Coleman, H., and Swenson, E. (1994). *DNA in the Courtroom: A Trial Watcher's Guide*. Seattle: Wadington Genelex.
Collins, F. (2000). Cited by C. Forbes. *The Australian*, June 28, p. 14.
Collins, P. H. (1990). *Black Feminist Thought*. London: Routledge.
Concar, D. (1995). Sex and the symmetrical body. *New Scientist* 146: 40–44.

Connell, R. W. (1987). *Gender and Power: Society, the Person and Sexual Politics*. Stanford: Stanford University Press.

Conrad, P., and Markens, S. (2001). Constructing the "gay gene" in the news: optimism and skepticism in the US and British press. *Health* 5: 373–400.

Cooke, B., Hegstrom, C. D., Villeneuve, L. S., and Breedlove, S. M. (1998). Sexual differentiation in the vertebrate brain: principles and mechanisms. *Frontiers in Neuroendocrinology* 19: 323–362.

Cooke, B. M., Tabibian, G., and Breedlove, S. M. (1999). A brain sexual dimorphism controlled by adult circulating androgens. *Proceedings of the National Academy of Sciences of the United States of America* 96: 7538–7540.

Cooley, C. H. (1922). *Social Processes*. New York: Charles Scribner's Sons.

Corbett, K. (2001). Faggot=loser. *Studies in Gender and Sexuality* 2: 3–28.

Cornell, D. (1992). Gender, sex and equivalent rights. In *Feminists Theorise the Political*, ed. J. Butler and J. W. Scott, pp. 280–296. New York: Routledge.

Cornell, D., and Thurschwell, A. (1987). Feminism, negativity, intersubjectivity. In *Feminism as Critique*, ed. S. Benhabib and D. Cornell, pp. 143–162. Cambridge: Polity.

Cosmides, L., and Tooby, J. (2000). Consider the source: the evolution of adaptations for decoupling and metarepresentation. In *Metarepresentations: A Multidisciplinary Perspective*, ed. D. Sperber pp. 53–115. New York: Oxford University Press.

Coward, R. (1984). *Female Desire: Women's Sexuality Today*. London: Granada.

——— (1985). Female desire and sexual identity. In *Women, Feminist Identity and Society in the 1980s*, ed. M. Diaz-Diocaretz and I. Zavala, n. p. Amsterdam: John Benjamins.

Crittenden, D. (2000). *What Our Mothers Didn't Tell Us: Why Happiness Eludes the Modern Woman*. New York: Touchstone Books.

Dalton, K. (1979). *Once a Month*. Sussex, UK: Harvester.

Dank, B. (1972). Why homosexual men marry women. *Medical Aspects of Human Sexuality* (August): 15–23.

Darrow, C. (1926). The eugenics cult. *American Mercury* 8: 129–137.

Darwin, C. (1871). *The Descent of Man*. London: John Murray.

Davidson, M. (1985). *Reach for the Top: A Woman's Guide to Success in Business and Management*. London: Piatkus.

Dawkins, R. (1976). *The Selfish Gene*. Oxford, UK: Oxford University Press.

de Lacoste, M.-C., Holloway, R. L., and Woodward, D. J. (1986). Sex differences in the fetal human corpus callosum. *Human Neurobiology* 5: 93–96.

de Lacoste-Utamsing, C., and Holloway, R. L. (1982). Sexual dimorphism in the human corpus callosum. *Science* 216: 1431–1432.

de Lepervanche, M. M. (1984). The "naturalness" of inequality. In *Ethnicity, Class and Gender in Australia*, ed. G. Bottomley & M. M. de Lepervanche, pp. 49–71. Sydney: Allen and Unwin.

de Waal, F. (1982). *Chimpanzee Politics: Power and Sex Among Apes*. London: Jonathan Cape.

de Waal, F., and Lanting, F. (1997). *Bonobo: The Forgotten Ape*. Berkeley: University of California Press.

Delgado, J. M. R. (1971). *Physical Control of the Mind: Towards a Psychocivilised Society*. New York: Harper & Row.

Denenberg, V. H. (1981). Hemispheric laterality in animals and the effects of early experience. *Behavioral Brain Sciences* 4: 1–49.

Denenberg, V. H., Berrebi, A. S., and Fitch, R. H. (1989). A factor analysis of the rat's corpus callosum. *Brain Research* 497: 271–279.

Denenberg, V. H., Fitch, R. H., Schrott, L. M., et al. (1991). Corpus callosum: interactive effects of infantile handling and testosterone in the rat. *Behavioral Neuroscience* 105: 562–566.

Deng, C., and Rogers, L. J. (2002). Social recognition and approach in the chick: lateralization and effect of visual experience. *Animal Behaviour* 63: 697–706.

Derrida, J. (1978a). *Spurs*, trans. B. Harlow. Chicago: University of Chicago Press.

——— (1978b). *Writing and Difference*, trans. A. Bass. London: Routledge & Kegan Paul.

Di Stefano, C. (1983). Masculinity as ideology in political theory: Hobbesian man considered. *Women's Studies International Forum* 6: 633–644.

——— (1991). *Configurations of Masculinity: A Feminist Perspective on Modern Political Theory*. Ithaca, NY: Cornell University Press.

Dickens, H. (1982). *What a Drag*. Sydney: Angus & Robertson.

Dickinson, R. L. (1948). The gynecology of homosexuality. In *Sex Variants: A Study of Homosexual Patterns*, ed. G. W. Henry, pp. 1069–1129. New York and London: Paul B. Hoeber.

Dietz, M. G. (1991). Hannah Arendt and feminist politics. In *Feminist Interpretations and Political Theory*, ed. M. L. Shanley and C. Pateman, pp. 232–252. London: Polity.

Disch, L. (1991). Towards a feminist conception of politics. *P.S.* (Washington, DC) 24: 501–504.

Dittmann, R. W., Kappes, M. E., and Kappes, M. H. (1992). Sexual behavior in adolescent and adult females with congenital adrenal hyperplasia. *Psychoneuroendocrinology* 17: 153–170.

Dörner, G. (1976a). Hormone-dependent brain development and behaviour. In *Hormones and Behaviour in Higher Vertebrates*, ed. J. Balthazart, E. Prove, and R. Gilles pp. 204–217. Berlin: Springer-Verlag.

——— (1976b). *Hormones and Brain Differentiation*. Amsterdam: Elsevier.

———(1979). Hormones and sexual differentiation of the brain. *Sex, Hormones and Behavior, Ciba Foundation Symposium* 62: 81–112.

Dörner, G., Geier, T., Ahrens, L., et al. (1980). Prenatal stress as possible aetiogenetic factor of homosexuality in human males. *Endokrinologie* 75: 365–386.

Dörner, G., Schenk, B., Schmiedel, B., and Ahrens, L. (1983). Stressful events in prenatal life of bi- and homosexual men. *Experimental and Clinical Endocrinology* 81: 83–87.

Duberman, M., Bauml, P., Vicinius, M., and Chauncey, G. Jr., eds. (1989). *Hidden from History: Reclaiming the Gay and Lesbian Past*. New York: New American Library.

Duchaine, B., Cosmides, L., and Tooby, J. (2001). Evolutionary psychology and the brain. *Current Opinion in Neurobiology* 11: 225–230.

Duster, T. (1990). *Backdoor to Eugenics*. New York: Routledge.

Eals, M., and Silverman, J. (1994). The hunter–gatherer theory of spatial sex differences: proximate factors mediating the female advantage in recall of object arrays. *Ethology and Sociobiology* 15: 95–105.

Eckert, E. D., Bouchard, T. J., Bohlen, J., and Heston, L. L. (1986). Homosexuality in monozygotic twins reared apart. *British Journal of Psychiatry* 14: 421–425.

Edut, O., ed. (1998). *Adiós, Barbie: Young Women Write about Body Image and Identity*. Seattle: Seal.

Ehrenreich, B., and English, D. (1976). *Witches, Midwives and Nurses: A History of Women Healers*. London: Writers and Readers.

Eisenstein, Z. R. (1988). *The Female Body and the Law*. Berkeley: University of California Press.

Elredge, N., and Tattersall, I. (1982). *The Myths of Human Evolution*. New York: Columbia University Press.

Elshtain, J. B. (1981). *Public Man, Private Woman: Women in Social and Political Thought*. Princeton, NJ: Princeton University Press.

——— (1986). *Meditations on Modern Political Thought: Masculine/Feminine Themes from Luther to Arendt*. New York: Praeger.

Engel, P. (1985). Androgynous zones. *Harvard Magazine* Jan–Feb: 24–32.

Ernst, E. (1956). *The Kabuki Theatre*. New York: Grove.

Evans, C. S. (1997). Referential signals. *Perspectives in Ethology* 12: 99–143.

Evans, D. T. (1993). *Sexual Citizenship: The Material Construction of Sexualities*. London: Routledge.

Eysenck, H. (1971). *The IQ Argument: Race, Intelligence and Education*. New York: Library.

Fabre-Nys, C. (1998). Steroid control of monoamines in relation to sexual behaviour. *Reviews of Reproduction* 3: 31–41.

Faludi, S. (1992). *Backlash: The Undeclared War Against Women*. London: Chatto & Windus.

Fausto-Sterling, A. (1992). *Myths of Gender: Biological Theories about Women and Men*. New York: Basic Books.

——— (2000). Beyond difference: feminism and evolutionary psychology. In *Alas Poor Darwin: Arguments Against Evolutionary Psychology*, ed. H. Rose and S. Rose, pp. 127–153. New York: Harmony Books.

——— (2001). Life in the XY corral. In *The Gender and Science Reader*, ed. M. Lederman and I. Bartsch, pp. 234–251. New York: Routledge.

Fee, E. (1979). Nineteenth-century craniology: the study of the female skull. *Bulletin of the History of Medicine* 53: 415–433.

Feingold, A. (1988). Cognitive gender differences are disappearing. *American Psychologist* 43: 95–103.

Fitch, R. H., and Denenberg, V. H. (1998). A role for ovarian hormones in sexual differentiation of the brain. *Behavioral and Brain Sciences* 21: 311–352.

Flynn, J. R. (1987). Massive IQ gains in 14 nations: what IQ tests really measure. *Psychological Bulletin* 101: 171–191.

——— (1994). IQ gains over time. In *Encyclopedia of Human Intelligence*. New York: Macmillan.

Fodor, J. (1983). *The Modularity of Mind*. Cambridge, MA: MIT Press.

Ford, C. S., and Beach, F. A. (1951). *Patterns of Sexual Behaviour*. New York: Harper & Row.

Foucault, M. (1977). *Discipline and Punish: The Birth of the Prison*. London: Allen Lane.

——— (1978). *A History of Sexuality*, trans. R. Hurley. New York: Pantheon.

——— (1985). *The Use of Pleasure*, vol. 2 of *The History of Sexuality*. New York: Pantheon.

Fraser, N. (1991). What's critical about critical theory: a case of Habermas and gender. In *Feminist Interpretations and Political Theory*, ed. M. L. Shanley and C. Pateman, pp. 253–276. London: Polity.

Frazer, J. G. (1974). *The Golden Bough: A Study in Magic and Religion*. London: Macmillan.

Friedman, L. (1995). The space factor in mathematics: gender differences. *Review of Educational Research* 65: 22–50.

Fuchs, J. A. P. (1983). On the war path and beyond: Hegel, Freud and feminist theory. *Women's Studies International Forum* 6: 565–572.

Gagnon, P. (1975). Sex research and social change. *Archives of Sexual Behavior* 4: 111–141.

Gallagher, S. A. (1989). Predictors of SAT mathematics score of gifted male and gifted female adolescents. *Psychology of Women Quarterly* 13: 191–203.

Gallup, G. G., Jr. (1982). Permanent breast enlargement in human females: a sociobiological analysis. *Journal of Human Evolution* 11: 597–601.

Galton, F. (1869). *Hereditary Genius*. London: Macmillan.

Gangestad, S. W., Thornhill, R., and Yeo, R. A. (1994). Facial attractiveness, developmental stability, and fluctuating asymmetry. *Ethology and Sociobiology* 15: 73–85.

Gatens, M. (1991). *Feminism and Philosophy: Perspectives on Difference and Equality*. London: Polity.

Gaulin, S. J. C., and Boster, J. S. (1992). Human marriage systems and sexual dimorphism in stature. *American Journal of Physical Anthropology* 89: 467–475.

Gaulin, S. J. C., and Fitzgerald, R. W. (1990). Sex differences in spatial ability and activity in two vole species (*Microtus ochrogaster* and *Microtus pennsylvanicus*). *Journal of Comparative Psychology* 104: 88–93.

Gaulin, S. J. C., and Sailer, L. D. (1984). Sexual dimorphism in weight amongst the primates: the relative impact of allometry and sexual selection. *International Journal of Primatology* 5: 515–535.

Gentry, S. E. (1993). Caring for lesbians in a homophobic society. In *Lesbian Health: What Are the Issues?*, ed. P. N. Stern, pp. 83–90. Indiana: Taylor and Francis.

Geschwind, N., and Galaburda, A. M. (1985). Cerebral lateralization, biological mechanisms, associations and pathology. *Archives of Neurology* 42: 634–654.

——— (1987). *Cerebral Lateralization: Biological Mechanisms, Associations, and Pathology*. Cambridge, MA: MIT Press.

Gilligan, C., Rogers, A. G., and Tolman, D., eds. (1991). *Women, Girls and Psychotherapy: Reframing Resistance*. New York: Harrington Park.

Ginsburg, F. D., and Rapp, R., eds. (1995). *Conceiving the New World Order: The Politics of Reproduction*. Berkeley: University of California Press.

Glickman, A. (1985). Androgyny is in at Harvard—sort of. *Harvard Magazine* Jan–Feb: 33.

Gonsiorek, J. C. (1995). Review of Hamer and Copeland's *The Science of Desire—The Search for the Gay Gene and the Biology of Behavior*. *Journal of Sex Research* 32: 262–263.

Gooch, G. P. (1956). *Louis XV. The Monarchy in Decline*. London.

Gordon, R. A. (1998). *Anorexia and Bulimia: Anatomy of a Social Epidemic*. Malden, MA: Blackwell.

Gosney, E., and Popenoe, P. (1929). *Sterilization for Human Betterment*. New York: Macmillan.

Gould, J. L., and Marler, P. (1987). Learning by instinct. *Scientific American* 256: 74–85.

Gould, S. J. (1981). *The Mismeasure of Man*. New York and London: Norton.

Goy, R. W., and Goldfoot, D. A. (1974). Experiential and hormonal factors influencing development of sexual behavior in the male rhesus monkey. In *The Neurosciences: Third Study Program*, ed. F. O. Schmitt and F. G. Worden, pp. 571–581. Cambridge, MA: MIT Press.

Grant, L., Ward, K. B., and Rong, X. L. (1987). Is there an association between gender and methods in sociological research? *American Sociological Review* 52: 856–862.

Gray, J. (1983). The married professional woman: an examination of her role conflicts and coping strategies. *Psychology of Women Quarterly* 7: 235–243.

Green, R. (1993). Dimensions of human sexual identity: transsexuals, homosexuals, fetishists, cross-gendered children and animal models. In *The Development of Sex Differences and Similarities in Behaviour*, ed. M. Haug, R. E. Whalen, C. Aron, and K. L. Olsen, pp. 477–486. Dordrecht, The Netherlands: Kluwer Academic Publishers.

——— (2002). Neurobiology of sexual identity and orientation. Paper presented at the Conference of the International Neuropsychoanalysis Society, Stockholm, Sept. 1–2, 2002.

Green, R., and Young, R. (2000). Fingerprint asymmetry in male and female transsexuals. *Personality and Individual Differences* 29: 933–942.

Greene, L. A., Töröcsik, B., Mendelsohn, C. L., et al. (2001). Gene regulation and neural development: ATF's and beyond. *International Journal of Developmental Neuroscience* 19: 685.

Greenfield, A., and Koopman, P. (1996). *SRY* and mammalian sex determination. *Current Topics in Developmental Biology* 34: 1–23.

Gross, E. (1986). Irigaray and sexual difference. *Australian Feminist Studies* 2: 63–78.

Grossman, A. (1983). The new woman and the rationalization of sexuality in Weimar Germany. In *Powers of Desire: the Politics of Sexuality*, ed. A. Snitow, C. Stansell, and S. Thompson, pp. 153–171. New York: Monthly Review.

Grosz, E. (1989). *Sexual Subversions*. Sydney: Allen and Unwin.

Guillaumin, C. (1985). The masculine: denotations/connotations. *Feminist Issues* 5(1): 65–74.

Guille-Escuret, G., and La Vigne, A. (1997). Biology reinvigorated: life/society, nature/culture, evolution/history. *Diogenes* 180: 1–20; also web: http://web5.infotrac.galegroup.com/itw/infomark/

Gunew, S., ed. (1990). *Feminist Knowledge: Critique and Construct*. New York: Routledge.

Gunew, S., and Yeatman, A., eds. (1993). *Feminism and the Politics of Difference*. Sydney: Allen and Unwin.

Gur, R. C., Mozley, L. H., Resnick, S. M., et al. (1995). Sex differences in regional cerebral glucose metabolism during resting state. *Science* 267: 528–531.

Habermas, J. (1971). *Knowledge and Human Interest*. Boston: Beacon.

Haier, R. J., and Benbow, C. P. (1995). Sex differences and lateralization in temporal lobe glucose metabolism during mathematical reasoning. *Developmental Neuropsychology* 11: 405–414.

Hall, R. L., ed. (1982). *Sexual Dimorphism in Homo Sapiens: A Question of Size*. New York: Praeger.

Halliday, F. E. (1964). *A Shakespeare Companion 1564–1964*. Harmondsworth, UK: Penguin.

Hamer, D. H., and Copeland, P. (1996). *The Science of Desire: The Search for the Gay Gene and the Biology of Behavior*. New York: Simon & Schuster.

——— (1998). *Living with Our Genes*. New York: Anchor Books.

Hamer, D. H., Hu, S., Magnuson, V. L., et al. (1993). A linkage between DNA markers on the X chromosome and male sexual orientation. *Science* 261: 321–327.

Hammer, M. (1977). *Hexenwahn und Hexenprozesse*. Frankfurt/Main: Fischer Verlag.

Hampson, E., and Kimura, D. (1992). Sex differences and hormonal influences on cognitive function in humans. In *Behavioral Endocrinology*, ed. J. B. Becker, S. M. Breedlove, and D. Crews, pp. 357–400. Cambridge, MA: MIT Press.

Hansell, N. K., Wright, M. J., Geffen, G. M., et al. (2001). Genetic influence on ERP slow wave measures of working memory. *Behavioral Genetics* 31: 603–614.

Haraway, D. J. (1982). Class, race, sex as scientific objects of knowledge: a Marxist-feminist perspective on the social construction of productive nature and some political consequences. *Argument* 24: 200–213.

——— (1985). A Manifesto for cyborgs: science, technology, and socialist feminism in the 1980s. *Socialist Review* 15: 65–108.

——— (1991). *Simians, Cyborgs and Women: The Reinvention of Nature*. New York: Routledge.

Harding, S., and O'Barr, J. eds. (1987). *Sex and Scientific Inquiry*. Chicago: University of Chicago Press.

Harper, J., and Heath, D. (1972). Separate and unequal: boys and girls at school. *Dissent* 28: 8–10.

Harré, R. (1981). The dramaturgy of sexual relations. In *The Bases of Human Sexual Attraction*, ed. M. Cook, pp. 251–274. London: Academic Press.

Harris, G. W., and Levine, S. (1965). Sexual differentiation of the brain and its experimental control. *Journal of Physiology* 181: 379–400.

Hart, J. (1992). A cocktail of alarm: same-sex couples and migration to Australia 1985–1990. In *Modern Homosexualities: Fragments of Lesbian and Gay Experiences*, ed. K. Plummer, pp. 121–133. London: Routledge.

——— (2002). *Stories of Gay and Lesbian Immigration: Together Forever?* New York and London: Haworth.

Haug, F., ed. (1987). *Female Sexualisation*, trans. E. Carter. London: Verso.

Heim, A. H. (1970). *Intelligence and Personality*. Harmondsworth, UK: Penguin.

Heller, A. (1977). *Instinkt, Aggression, Charakter*. Berlin: VSA.

——— (1982). The emotional division of labour between the sexes: perspectives on feminism and socialism. *Thesis Eleven* 5–6: 59–73.

Hellige, J. B. (1993). *Hemispheric Asymmetry: What's Right and What's Left*. Cambridge, MA: Harvard University Press.

Henry, G. W. (1948). *Sex Variants: A Study of Homosexual Patterns*. New York: Hoeber.

Henry, W. A., III (1993). Born gay? *Time* July 26, pp. 44–47.

Herbison, A. E., and Theodosis, D. T. (1992). Immunocytochemical identification of oestrogen receptors in preoptic neurones containing calictonin gene-regulated peptide in the male and female rat. *Neuroendocrinology* 56: 761–764.

Herdt, G. H. (1993). *Ritualized Homosexualities in Melanesia*. Berkeley: University of California Press.

Heriot, A. (1956). *The Castrati in Opera*. London: Secker & Warburg.

Herrnstein, R. J., and Murray, C. (1994). *The Bell Curve. Intelligence and Class Structure in American Life*. New York: The Free Press.

Hines, M. (1993). Hormonal and neural correlates of sex-typed behavioral development in human beings. In *The Development of Sex Differences and Similarities in Behaviour*, ed. M. Haug, R. E. Whalen, C. Aron, and K. L. Olsen, pp. 131–149. Dordrecht, The Netherlands: Kluwer Academic Publishers.

——— (1998). The trouble with sex. *Nature* 392:146.

Hirschfeld, L. A., and Gelman, S. A., eds. (1994). *Mapping the Mind: Domain Specificity in Cognition and Culture*. Cambridge, UK: Cambridge University Press.

Hoagland, S. L. (1982). Femininity, resistance and sabotage. In *Femininity, Masculinity and Androgyny*, ed. M. Vettering-Braggin, pp. 85–98. Totowa, NJ: Littlefield, Adams.

Hobbes, T. (1955) *Leviathan*, ed. M. Oakeshott. Oxford, UK: Basil Blackwell.

Hoch, P. (1979). *White Hero, Black Beast: Racism, Sexism, and the Mask of Masculinity*. London: Pluto Press.

Hoffman, C., and Hurst, N. (1990). Gender stereotypes: Perception or rationalization? *Journal of Personality and Social Psychology* 58: 197–208.

Holcomb, H. R., III, ed. (2001). *Conceptual Challenges in Evolutionary Psychology: Innovative Research Strategies*. New York: Kluwer Academic Publishers.

Honig, B. (1992). Toward an agonistic feminism: Hannah Arendt and the politics of identity. In *Feminists Theorize the Political*, ed. J. Butler and J. W. Scott, pp. 215–238. New York: Routledge.

Hoogland, R. C. (1999). First things first: Freud and the question of primacy in gendered sexuality. *Journal of Gender Studies* 8: 43–56.

hooks, B. (1981). *Ain't I a Woman: Black Women and Feminism*. Boston: South End.

——— (2000). *Feminism Is for Everybody: Passionate Politics*. Boston: South End.

Horgan, J. (1993). Eugenics revisited: trends in behavioral genetics. *Scientific American* 268: 92–100.

Hoult, T. F. (1983/1984). Human sexuality in biological perspective: theoretical and methodological considerations. *Journal of Homosexuality* 9: 2–3, 137–155.

Hu, S., Pattatucci, A. M. L., Patterson, C., et al. (1995). Linkage between sexual orientation and chromosome Xq28 in males but not females. *Nature Genetics* 11: 248–256.

Hubbard, R. (1994). Race and sex as biological categories. In *Challenging Racism and Sexism: Genes and Gender, VII*, ed. E. Tobach and B. Rosoff, pp. 11–21. New York: Feminist Press at the City University of New York.

Hubbard, R., and Wald, E. (1993). *Exploding the Gene Myth.* Boston: Beacon.

Hutchison, J. B., and Steimer, T. L. (1984). Androgen metabolism in the brain: behavioural correlates. In *Sex Differences in the Brain: Progress in Brain Research*, ed. G. L. De Vries, J. P. C. de Bruin, H. B. M. Ujlings, and M. A. Corner, 61: 23–51.

Hutt, C. (1972). *Males and Females.* Harmondsworth, UK: Penguin.

Hyde, J. S., Fennema, E., and Lamon, S. J. (1990). Gender differences in mathematics performance: a meta-analysis. *Psychology Bulletin* 107: 139–155.

Hyslop, G. (1985). Deviant and dangerous behavior. *Journal of Popular Culture* 19: 65–77.

Illich, I. (1977). *Limits to Medicine: Medical Nemesis: the Expropriation of Health.* Harmondsworth, UK: Penguin.

Imielinski, K. (1969). Homosexuality in males with particular reference to marriage. *Psychotherapy and Psychosomatics* 17: 126–132.

Irigaray, L. (1974). *Speculum*, trans. G. Gill. Paris: Minuit, 1985. *Speculum of the Other Woman.* Ithaca, NY: Cornell University Press.

——— (1977). *Ce sexe qui n'en est pas un.* Paris: Minuit; trans. C. Porter (1985). *This Sex Which Is Not One.* Ithaca, NY: Cornell University Press.

——— (1984). *Éthique de la différence sexuelle.* Paris: Editions de Minuit.

Jacobs, J. E., and Eccles, J. S. (1992). The impact of mother's gender-role stereotypic beliefs on mother's and children's ability perceptions. *Journal of Personality and Social Psychology* 63: 932–944.

Jacobs, L. F., and Spencer, W. D. (1994). Natural space-use patterns and hippocampal size in kangaroo rats. *Brain, Behavior, and Evolution* 44: 125–132.

Jeffreys, S. (1993). *The Lesbian Heresy: Feminist Perspective in the Lesbian Sexual Revolution.* Melbourne: Spinifex.

Jennett, C., and Stewart, R. G., eds. (1987). *Three Worlds of Inequality: Race, Class and Gender*. Melbourne: Macmillan.

Jensen, A. R. (1969). How much can we boost IQ and scholastic achievement? *Harvard Education Revue* 39: 1–123.

Jolly, M. (1978). The politics of difference: feminism, colonialism and decolonisation in Vanuatu. In *Intersexions: Gender/Class/Culture/Ethnicity*, ed. G. Bottomley, M. de Lepervanche, and J. Martin. Sydney: Allen & Unwin. (Speech by Grace Mera Molisa to the First Conference of Vanuaaku Women, Efate, Vanuatu, 1978, cited on p. 52.)

Jussim, L., and Eccles, J. S. (1992). Teacher expectations II: construction and reflection of student achievement. *Journal of Personality and Social Psychology* 63: 947–961.

Kalick, S. M., Zebrowitz, L. A., Langlois, J. H., and Johnson, R. M. (1998). Does human facial attractiveness honestly advertise health? *Psychological Science* 9: 8–13.

Kamenka, E., and Erh-Soon, A. T., eds. (1978). *Human Rights*. London: Edward Arnold.

Kamin, L. J. (1977). *The Science & Politics of I.Q.* Harmondsworth, UK: Penguin.

Kaplan, G. (1992). *Contemporary Western European Feminism*. Sydney: Allen and Unwin; London: UCL; New York: New York University Press.

——— (1993). Accounting for difference: a review of feminism and political theory. *Political Theory Newsletter* 5: 140–163.

——— (1994). Irreducible "human nature": Nazi views on Jews and women. In *Challenging Racism and Sexism: Alternatives to Genetic Determinism, Genes and Gender*, VII, ed. E. Tobach and B. Rosoff, pp. 188–211. New York: Feminist Press at the City University of New York.

——— (1996). *The Meager Harvest: The Australian Women's Movement 1950s to 1990s*. Sydney: Allen & Unwin.

——— (1999). Pluralism and citizenship: The case of gender in European politics. In *Citizenship and National Identity in Europe*, ed. P. Murray and L. Holmes, pp. 73–95. London: Ashgate.

——— (2000a). Women and development: Western Europe. *Routledge International Encyclopedia of Women: Global Women's Issues and*

Knowledge, eds. C. Kramarae and D. Spender, vol. 1, pp. 369–375. New York: Routledge.

——— (2000b). European respectability, eugenics and globalization. In "*A Race for a Place*": *Eugenics, Darwinism and Social Thought and Practice in Australia*, ed. M. Crotty, J. Germov, and G. Rodwell. Proceedings of the History & Sociology of Eugenics Conference, University of Newcastle, 27–28 April 2000, pp. 19–26. The University of Newcastle, Newcastle, Australia.

——— (2001). New social movements. *Encyclopedia of Modern European Social History*, vol 3, pp. 289–300. New York: Charles Scribner's Sons.

Kaplan, G., and Adams, C. E. (1990). Early women supporters of national socialism. In *The Attractions of Fascism*, ed. J. Milfull, pp. 186–203. Hamburg: Berg.

Kaplan, G., and Kessler, C. S., eds. (1989). *Hannah Arendt: Thinking, Judging, Freedom: Collected Essays*. Boston, London, and Sydney: Allen and Unwin.

Kaplan, G., and Rogers, L. J. (1984). Breaking out of a dominant paradigm: a new look at sexual attraction. *Journal of Homosexuality* 10: 71–75.

——— (1990). Scientific construction, cultural productions: scientific narratives of sexual attraction. In *Feminine/Masculine and Representation*, ed. A. Cranny-Francis and T. Threadgold, pp. 211–230. Sydney: Allen and Unwin.

——— (1994). Race and gender fallacies: the paucity of biological determinist explanations of difference. In *Challenging Racism and Sexism*, ed. E. Tobach and B. Rosoff, pp. 66–92. New York: The Feminist Press.

——— (2000). *The Orang-Utans: Their Evolution, Behavior, and Future*. Cambridge, MA: Perseus.

——— (2001). *Birds: Their Habits and Skills*. Sydney: Allen and Unwin.

Kaplan-Solms, K., and Solms, M. (2002). *Clinical Studies in Neuro-Psychoanalysis: Introduction to a Depth Neuropsychology*, 2nd ed. New York: Karnac.

Karkau, K. (1976). Sexism in the fourth grade. In *Undoing Sex Stereotypes: Research and Resources for Educators*, ed. M. Guttentag and H. Bray, pp. 64–80. New York: McGraw-Hill.

Karmiloff-Smith, A. (2000). Why babies' brains are not Swiss army knives. In *Alas Poor Darwin: Arguments Against Evolutionary Psychology*, ed. H. Rose and S. Rose, pp. 173–187. New York: Harmony Books.

Keller, E. F. (1983). *A Feeling for the Organism: The Life and Work of Barbara McClintock*. New York: Freeman.

——— (1985). *Reflections on Gender and Science*. New Haven, CT: Yale University Press.

Kennedy, R., and Coonan, H. (1975). Civil liberties and the lesbian. *Female Homosexuality*. *Camp Ink*, Sydney: University of Sydney: 34–44.

Kertesz, A., and Benke, T. (1989). Sex equality in hemispheric language organization. *Brain and Language* 37: 401–408.

Kimura, D. (1983). Sex differences in cerebral organisation for speech and praxic functions. *Canadian Journal of Psychology* 27: 19–25.

——— (1992). Sex differences in the brain. *Scientific American* 267: 81–87.

——— (1999). *Sex and Cognition*. London: A Bradford Book; Cambridge, MA: MIT Press.

Kinsbourne, M. (1980). If sex differences in the brain exist, they have yet to be discovered. *The Behavioral and Brain Sciences* 3: 241–242.

Kinsey, A. C., Pomeroy, W. B., and Martin, C. E. (1948a). *Sexual Behavior in the Human Female*. Philadelphia: Saunders.

——— (1948b). *Sexual Behavior in the Human Male*. Philadelphia: Saunders.

Kowner, R. (1994). The effect of physical attractiveness comparison on choice of partners. *The Journal of Social Psychology* 135(2): 153–165.

Kramer, O. P. (1994). *Listening to Prozac*. New York: Viking Penguin.

Kurth, G. (1967). Implications of primate paleontology for behavior. In *Genetic Diversity and Human Behavior*, ed. J. N. Spuhler, n.p. Chicago: University of Chicago Press.

Kurzban, R. (2002). Alas poor evolutionary psychology: unfairly accused, unjustly condemned: essay review (*Alas*, *Poor Darwin*, ed. H. and R. Rose). *The Human Nature Review* 2: 99–109.

Lai, C. S. L., Fisher, S. E., Hurst, J. A., et al. (2001). A fork-head domain gene is mutated in a severe speech and language disorder. *Nature* 413: 519–523.

Lambert, H. H. (1978). Biology and equality. *Signs* 4: 97–117.

Laplanche, J., and Pontalis, J.-B. (1988). *The Language of Psychoanalysis.* London: Karnac and the Institute of Psycho-Analysis.

Larsson, K. (1978). Experiential factors in the development of sexual behaviour. In *Biological Determinants of Sexual Behaviour*, ed. J. B. Hutchison, pp. 55–86. Chichester, UK: Wiley.

Lather, P. (1991). *Getting Smart: Feminist Research and Pedagogy With/In the Postmodern.* New York: Routledge.

Leacock, E. (1986). The anthropology of war: biological determinism in cultural clothing. Paper delivered at *Genes and Gender Conference.* Stern College, New York. (See Casey for discussion.)

LeVay, S. (1991). A difference in hypothalamic structure between heterosexual and homosexual men. *Science* 253: 1034–1037.

——— (1993). *The Sexual Brain.* Cambridge, MA: MIT Press.

Leakey, R. (1994). *The Origin of Humankind.* New York: Basic Books.

Lennon, M. C., and Rosenfield, S. (1994). Relative fairness and the division of housework: the importance of options. *American Journal of Sociology* 100: 506–531.

Levinson, D. F., Holmans, P. A., Laurent, C., et al. (2002). No major schizophrenia locus detected on chromosome 1q in a large multicenter sample. *Science* 296: 739–741.

Levy, J. (1977). The mammalian brain and the adaptive advantage of cerebral asymmetry. *Annals of the New York Academy of Science* 229: 265–272.

Levy, J., and Gurr, R. C. (1980). Individual differences in psychoneurological organization. In *Neuropsychology of Left-Handedness*, ed. J. Herran, pp. 199–210. New York: Academic Press.

Lewins, F. (1995). *Transsexualism in Society: A Sociology of Male-to-Female Transsexuals.* Melbourne: Macmillan Education.

Lewontin, R. C. (1991). *The Doctrine of DNA.* London: Penguin.

Lewontin, R. C., Rose, S., and Kamin, L. J. (1984). *Not in Our Genes: Biology, Ideology, and Human Nature.* New York: Pantheon.

Li, W.-H., and Makova, K. (2002). Strong male-driven evolution of DNA sequences in humans and apes. *Nature* 416: 624–626.

Libby, M. S. (1935). *The Attitude of Voltaire to Magic and the Sciences.* New York: n.p.

Lidz, T. (1993). Reply to "A Genetic Study of Male Sexual Orientation." *Archives of General Psychiatry* 50: 240.

Lloyd, G. (1984). *The Man of Reason: "Male" and "Female" in Western Philosophy*. London: Methuen.

——— (1991). Reason as attainment. In *Reader in Feminist Knowledge*, ed. S. Gunew, pp. 166–180. New York: Routledge.

Lorber, J. (1999). Embattled terrain: gender and sexuality. In *Revisioning Gender*, ed. M. M. Ferree, J. Lorber, and B. B. Hess, pp. 416–448. Thousand Oaks, CA: Sage.

Love, N. S. (1991). Politics and voice(s): an empowerment/knowledge regime. *Differences: A Journal of Feminist Cultural Studies* 3(1): 85–103.

Luciano, M., Wright, M. J., Smith, G. A., et al. (2001). Genetic covariance among measures of information processing speed, working memory, and IQ. *Behavioral Genetics* 31:581–592.

Lumby, C. (1997). *Bad Girls: The Media, Sex and Feminism in the '90s*. Sydney: Allen and Unwin.

Lumsden, C. J., and Wilson, E. O. (1981). *Genes, Mind and Culture: The Coevolutionary Process*. Cambridge, MA: Harvard University Press.

Lytton, H., and Romney, D. M. (1991). Parents' differential socialization of boys and girls: a meta-analysis. *Psychological Bulletin* 109: 267–296.

Mackerras, C. P. (1972). *The Rise of the Peking Opera 1770–1870*. Oxford, UK: Clarendon.

Mackintosh, N. J. (2000). Evolutionary psychology meets *g*. *Nature* 403: 378–379.

Mall, F. P. (1909). On several anatomical characters of the human brain, said to vary according to race and sex, with special reference to the weight of the frontal lobe. *American Journal of Anatomy* 9: 1–32.

Mandler, G. (1997). *Human Nature Explored*. New York: Oxford University Press.

Marcuse, H. (1968). *Marcuse Negations*, trans. J. J. Schapiro. Harmondsworth: Penguin.

Mark, V. H., and Ervin, F. R. (1970). *Violence and the Brain*. New York: Harper and Row.

Marks, J. (2002). *What It Means to Be 98% Chimpanzee: Apes, People, and Their Genes*. Berkeley: University of California Press.

Marshall, E. (1995). NIH's "Gay Gene" study questioned. *Science* 268: 1841.

Martin, C. L., Eisenbud, L., and Rose, H. (1995). Children's gender-based reasoning about toys. *Child Development* 66: 1453–1471.

Martin, E. (1991). The egg and the sperm: how science has constructed a romance based on stereotypical male-female roles. *Signs* 16(3): 485–501.

Masters, M. S., and Sanders, B. (1993). Is the gender difference in mental rotation disappearing? *Behavioral Genetics* 23: 337–342.

Matthews, J. J. (1984). *Good and Mad Women*. Sydney: Allen and Unwin.

Matthews, W. C., Booth, M. W., Turner, J. D., and Kessler, L. (1986). Physicians' attitudes towards homosexuality: survey of a Californian county medical society. *Western Journal of Medicine* 144: 106–110.

McBurney, D. H., Gaulin, S. C. G., Devineni, T., and Adams, C. (1997). Superior spatial memory of women: stronger evidence for the gathering hypothesis. *Evolution and Human Behavior* 18: 165–174.

McCormick, C. M., Witelson, S. F., and Kingstone, E. (1990). Left-handedness in homosexual men and women: neuroendocrine implications. *Psychoneuroendocrinology* 15: 69–76.

McGue, M., and Bouchard, T. J. (1998). Genetic and environmental influences on human behavioral differences. *Annual Reviews in Neuroscience* 21: 1–24.

McKeever, W. F. (1995). Hormone and hemisphericity hypothesis regarding cognitive sex differences: possible future explanatory power, but current empirical chaos. *Learning and Individual Differences* 7: 323–340.

McKinney, F. S., Derrickson, R., and Mineau, P. (1983). Forced copulation in waterfowl. *Behaviour* 86: 250–294.

Mead, G. H. (1934). *Mind, Self and Society*. Chicago: University of Chicago Press.

Mead, M. (1935). *Sex and Temperament in Three Primitive Societies*. New York: William Morrow, 1988.

——— (1950). *Male and Female: A Study of the Sexes in a Changing World*. Harmondsworth, UK: Penguin, 1974.

Melucci, A. (1988). Social movements and the democratization of everyday life. In *Civil Society and the State*, ed. J. Keane, pp. 245–260. London/New York: Verso.

Mezey, S. G. (1990). When should differences mark difference?: A new approach to the constitutionality of gender based laws. *Women and Politics* 10: 105–119.

Mill, J. S (1869). *The Subjection of Women*. Philadelphia: Lippincott.
Miller, E. M. (1997). *Recent Explorations in Biology and Politics*. JAI Press.
Miller, G. F. (1988). How mate choice shaped human nature: a review of sexual selection and human evolution. In *Handbook of Evolutionary Psychology: Ideas, Issues, and Applications*, ed. C. Crawford and D. Krebs, pp. 87–129. Mahwah, N.J.: Erlbaum.
Miller, S. K. (1993). Gene hunters sound warning over gay link. *New Scientist* July 24, pp. 4–5.
Millett, K. (1977). *Sexual Politics*. London: Virago.
Mirsky, S. (1999). Supply and demand. *Scientific American* 280(5): 13.
Mirza, H. S. (1992). *Young, Female and Black*. London: Routledge.
Mitchell, J. (1966). Women: the longest revolution. *New Left Review* 40: 11–37.
Mithen, S. (1996). *The Prehistory of the Mind: The Cognitive Origins of Art, Religion, and Science*. New York: Thames and Hudson.
Moir, A., and Jessel, D. (1993). *Brainsex: The Real Difference Between Men and Women*. New York: Dell.
Money, J., and Dalery, J. (1976). Iatrogenic homosexuality. *Journal of Homosexuality* 1: 357–371.
Money, J., and Ehrhardt, A. A. (1967). Progestin-induced hermaphroditism: I.Q. and psychosexual identity in the study of ten girls. *Journal of Sex Research* 3:83–100.
——— (1972). *Man and Woman; Boy and Girl: The Differentiation and Dimorphism of Gender Identity from Conception to Maturity*. Baltimore, MD: Johns Hopkins University Press.
Moore, C. L. (1981). An olfactory basis for maternal discrimination of sex of offspring in rats. *Animal Behaviour* 29: 383–386.
——— (1982). Maternal behaviour of rats is affected by hormonal condition of pups. *Journal of Comparative and Physiological Psychology* 96: 123–129.
——— (1984). Maternal contributions to the development of masculine sexual behavior in laboratory rats. *Developmental Psychobiology* 17: 347–356.
——— (1985a). Sex differences in urinary odors produced by young laboratory rats. *Journal of Comparative Psychology* 99: 336–341.
——— (1985b). Another psychobiological review of sexual differentiation. *Developmental Review* 5: 18–55.
——— (1995). Maternal contributions to mammalian reproductive de-

velopment and the divergences of males and females. *Advances in the Study of Behavior* 24:47–118.

Moore, C. L., Dou, H., and Juraska, J. M. (1992). Maternal stimulation affects the number of motor neurons in a sexually dimorphic nucleus of the lumbar spinal cord. *Brain Research* 572: 52–56.

——— (1996). Number, size, and regional distribution of motor neurons in the dorsolateral and retrodorsal nuclei as a function of sex and neonatal stimulation. *Developmental Psychobiology* 29: 303–313.

Moore, C. L., and Morelli, G. A. (1979). Mother rats interact differently with male and female offspring. *Journal of Comparative and Physiological Psychology* 93: 677–684.

Moore, C. L., Wong, L., Daum, M. C., and Leclair, O. U. (1997). Mother–infant interactions in two strains of rats: implications for dissociating mechanism and function of a maternal pattern. *Developmental Psychobiology* 30: 301–312.

Morris, D. (1967). *The Naked Ape*. New York: McGraw-Hill.

Mosse, G. L. (1978). *Towards the Final Solution: A History of European Racism*. New York: Howard Fertig.

Mouffe, C. (1992). Feminism, citizenship and radical democratic politics. In *Feminists Theorize the Political*, ed. J. Butler and J. W. Scott, pp. 369–384. New York: Routledge.

Mueller, C. M., ed. (1988). *The Politics of the Gender Gap: The Social Construction of Political Influence*. Newbury Park, CA: Sage.

Mueller, U., and Mazur, A. (1996). Facial dominance of West Point cadets as a predictor of later military rank. *Social Forces* 74: 823–850.

Mulford, M., Orbell, J., Shatto, C., and Stockard, J. (1998). Physical attractiveness, opportunity, and success in everyday exchange. *American Journal of Sociology* 103: 1565–1592.

Nachman, M. W., Brown, W. M., Stoneking, M., and Aquadro, C. F. (1996). Nonneutral mitochondrial DNA variation in humans and chimpanzees. *Genetics* 142: 953–963.

Nadler, R. D. (1995). Sexual behavior of orangutans: basic and applied implications. In *The Neglected Ape*, ed. R. D. Nadler, B. F. M. Galdikas, L. K. Sheeran, and N. Rosen, pp. 223–237. New York: Plenum.

Nakamichi, M., Cho, F., and Minami, T. (1990). Mother–infant interactions of wild-born, individually-caged cynomolgus monkeys (*Macaca fascicularis*) during the first 14 weeks of infant life. *Primates* 31: 213–224.

Nakano, G. E. (1999). The social construction and institutionalization of gender and race: an integrative framework. In *Revisioning Gender*, ed. M. M. Ferree, J. Lorber, and B. B. Hess, pp. 3–43. Thousand Oaks, CA: Sage.

Nelkin, D. (2000). Less selfish than sacred?: Genes and the religious impulse in evolutionary psychology. In *Alas Poor Darwin: Arguments Against Evolutionary Psychology*, ed. H. Rose and S. Rose, pp. 17–32. New York: Harmony Books.

Neumann, E. (1954). *The Origins and History of Consciousness*. New York: Pantheon.

Nordeen, E. J., and Yahr, P. (1982). Hemispheric asymmetries in the behavioral and hormonal effects of sexually differentiating mammalian brain. *Science* 218: 391–394.

Nuffield Council on Bioethics (2002). *Genetics and Human Behaviour: The Ethical Context*. Nuffield Council, UK. Also at website www.nuffield bioethicsorg/filelibrary/pdf/nuffieldgeneticsrep.pdf.

Oakley, M. A. (1972). *Elizabeth Cady Stanton*. New York: Feminist Press.

O'Brien, M. (1989). *Reproducing the World: Essays in Feminist Theory*. Boulder, CO: Westview.

Okin, S. M. (1980). *Women in Western Political Thought*. London: Virago.

O'Leary, J., and Sharp, R., eds. (1991). *Inequalities in Australia: Slicing the Cake*. Melbourne: Heinemann.

Olsen, K. L. (1983). Genetic determinants of sexual differentiation. In *Hormones and Behaviour in Higher Vertebrates*, ed. J. Balthazart, E. Prove, and R. Gilles, pp. 138–158. Heidelberg, New York, and Toronto: Springer-Verlag.

Oyama, S. (1985). *The Ontogeny of Information: Developmental Systems and Evolution*. Cambridge, UK: Cambridge University Press.

——— (1993). Constraints on development. *Netherlands Journal of Zoology* 43: 6–16.

Page, D. C., Mosher, R., Simpson, E. M., et al. (1987). The sex-determining region of the human Y chromosome encodes a finger protein. *Cell* 51: 1091–1104.

Pakkenberg, B., and Gundersen, H. J. (1997). Neocortical neuron number in humans: effect of sex and age. *Journal of Comparative Neurology* 384: 312–320.

Palmer, J. A., and Palmer, L. K. (2002). *Evolutionary Psychology: The Ultimate Origins of Human Behavior*. Boston: Allyn and Bacon.

Parlee, M. R. (1973). The premenstrual syndrome. *Psychological Bulletin* 80: 454.

Pasquinelli, C. (1984). Beyond the longest revolution: the impact of the Italian women's movement on cultural and social change. *Praxis International* 4(2): 131–136.

Pateman, C. (1988). *The Sexual Contract*. Stanford, CA: Stanford University Press.

——— (1991). God hath ordained to man a helper: Hobbes, patriarchy and conjugal right. In *Feminist Interpretations and Political Theory*, ed. M. L. Shanley and C. Pateman, pp. 53–73. London: Polity Press.

Paunio, T., Ekelund, J., Varilo, T., et al. (2001). Genome-wide scan in a nationwide study sample of schizophrenia families in Finland reveals susceptibility loci on chromosomes 2q and 5q. *Human Molecular Genetics* 10: 3037–3048.

Pawlowski, B. (1999). Loss of oestrus and concealed ovulation in human evolution. *Current Anthropology* 40(3): 257–275.

Pearson, K. (1901). *National Life from the Standpoint of Science*. London: A. and C. Black.

Peel, J. D. Y. (1972). *Herbert Spencer on Social Evolution*. Chicago: University of Chicago Press.

Peritore, N. P. (1997). The evolutionary critique of culturalism. *Research in Biopolitics* 5: 109–129.

Petrie, B. (1987). Angry women. In *Up From Below: Poems of the 1980s*, ed. I. Coates, N. J. Lorbett, and B. Petrie, p. 268. Sydney: Women's Redress Press.

Petrie, M., Halliday, T., and Sanders, C. (1991). Peahens prefer peacocks with elaborate trains. *Animal Behavior* 41: 323–331.

Phillips, G., and Over, R. (1995). Differences between heterosexual, bisexual, and lesbian women in recalled childhood experiences. *Archives of Sexual Behavior* 24: 1–20.

Phoenix, C. H., Goy, R. W., Gerall, A. A., and Young, W. C. (1959). Organizing action of prenatally administered testosterone propionate on the tissues mediating mating behaviour in the guinea pig. *Endocrinology* 65: 369–382.

Phoenix, C. H., Goy, R. W., and Resko, J. A. (1968). Psychosexual differentiation as a function of androgenic stimulation. In *Perspectives in Reproduction and Sexual Behavior*, ed. H. Diamond, pp. 33–49. Bloomington, IN: Indiana University Press.

Pigliucci, M., and Kaplan, J. (2000). The fall and rise of Dr. Pangloss: adaptationism and the *Spandrels* paper 20 years later. *Trends in Evolutionary Biology* 15: 66–70.

Pillard, R. C., and Bailey, J. M. (1995). A biological perspective on sexual orientation. *The Psychiatric Clinics of North America* 18: 71–84.

——— (1998). Human sexual orientation has a heritable component. *Human Biology* 70: 347–365.

Pinker, S. (1994). *The Language Instinct*. New York: William Morrow.

——— (1997). *How the Mind Works*. New York: Norton.

——— (2001). Talk of genetics and vice versa. *Nature* 413: 465–466.

Pitkin, H. F. (1984). *Fortune is a Woman: Gender and Politics in the Thought of Niccolo Machiavelli*. Berkeley: University of California Press.

Plato (1951). *The Symposium*, transl. W. Hamilton. London: Penguin.

Plummer, D. (1999). *One of the Boys: Masculinity, Homophobia and Modern Manhood*. New York: Haworth.

Plummer, K., ed. (1992). *Modern Homosexualities: Fragments of Lesbian and Gay Experiences*. London: Routledge.

Pollitt, K. (1999). The Solipsisters. *New York Times Book Review*, April 18, p. 35.

Pool, R. (1993). Evidence for homosexuality gene. *Science* 261: 291–292.

Poole, M., and Langen-Fox, J. (1997). *Australian Women and Careers: Psychological and Contextual Influences over the Lifecourse*. Melbourne: Cambridge University Press.

Posthuma, D., De Geus, E. J. C., and Boomsma, D. I. (2001). Perceptual speed and IQ are associated through common genetic factors. *Behavioral Genetics* 31: 593–602.

Probyn, E. (1988). The anorexic body. In *Body Invaders: Sexuality and the Postmodern Condition*, ed. A. and M. Kroker, pp. 201–212. London: Macmillan.

Pruner, F. (1866). Article in *Transactions of the Ethnological Society* 4: 13–33. Quoted by E. Fee (1979). Nineteenth-century Craniology: The Study of the Female Skull. *Bulletin of the History of Medicine* 53: 415–433.

Pugh, K. R., Shaywitz, B. A., Shaywitz, S. E., et al. (1996). Cerebral organisation of component processes in reading. *Brain* 119: 1221–1238.

Raag, T., and Rackliff, C. L. (1998). Preschoolers' awareness of social expectations of gender relationships to toy choices. *Sex Roles* 38: 685–700.

Ralls, K. (1976). Mammals in which females are larger than males. *Quarterly Review of Biology* 51: 245–276.

Randall, C. E. (1989). Lesbian phobia among BSN educators: a survey. *Journal of Nursing Education* 28: 302–306.

Rech, J. F. (1996). Gender differences in mathematics achievement and other variables among university students. *Journal of Research and Development in Education* 29: 73–76.

Reinisch, J. M., Ziemba-Davis, M., and Sanders, S. A. (1991). Hormonal contributions to sexually dimorphic behavioral development in humans. *Psychoneuroendocrinology* 16: 213–278.

Rhodes, D. L. (1989). *Justice and Gender: Sex Discrimination and the Law.* Cambridge, MA: Harvard University Press.

Rhodes, G., Hickford, C., and Jeffery, L. (2000). Sex-typicality and attractiveness: Are supermale and superfemale faces super-attractive? *British Journal of Psychology* 91: 125–140.

Rice, G., Anderson, C., Risch, N., and Ebers, G. (1999). Male homosexuality: absence of linkage to microsatellite markers at xq28. *Science* 284: 665–667.

Rich, A. (1976). *Of Woman Born.* New York: Norton.

——— (1980). Compulsory heterosexuality and lesbian existence. *Signs* 5(4): 631–660.

Richards, E. (1983). Darwin and the descent of woman. In *The Wider Domain of Evolutionary Thought*, ed. D. Olroyd and I. Langham, pp. 57–111. Dordrecht: Reidel.

Ridley, M. (1994). *The Red Queen: Sex and the Evolution of Human Nature.* New York: Penguin.

Riger, S. (1992). Epistemological debates, feminist voices: science, social values, and the study of women. *American Psychologist* 47: 730–740.

Roberts, B. (1983). Invisible people: homosexuals in Australia. *Meanjin* 42(4): 486–490.

Rodman, P. S., and Mitani, J. C. (1987). Orangutans: sexual dimorphism in a solitary species. In *Primate Societies*, ed. B. B. Smuts, D. L. Cheney, R. M. Seyfarth, et al., pp. 146–154. Chicago: University of Chicago Press.

Rogers, A. R., and Mukherjee, A. (1992). Quantitative genetics of sexual dimorphism in human body size. *Evolution* 46: 226–234.

Rogers, L. J. (1979). Menstruation. *Australian Family Physician* 8(8): 923–929.

——— (1981). Biology: gender differentiation and sexual variation. In *Australian Women: Feminist Perspectives*, ed. N. Grieve and P. Grimshaw, pp. 44–57. Melbourne: Oxford University Press.

——— (1982). Light experience and asymmetry of brain function in chickens. *Nature* 297: 223–225.

——— (1988). Biology, the popular weapon: sex differences in cognitive function. In *Crossing Boundaries: Feminisms and the Critique of Knowledges*, ed. B. Caine, E. Grosz, and M. de Lepervanche, pp. 43–51. Sydney: Allen and Unwin.

——— (1997). *Minds of Their Own*. Boulder, CO: Westview.

——— (2001). *Sexing the Brain*. New York: Columbia University Press.

Rogers, L. J., and Andrew, R. J., eds. (2002). *Comparative Vertebrate Lateralization*. Cambridge, UK: Cambridge University Press.

Rogers, L. J., and Anson, J. M. (1979). Lateralization of function in the chicken forebrain. *Pharmacology, Biochemistry and Behavior* 10: 679–686.

Rogers, L. J., and Kaplan, G. (2000). *Songs, Roars and Rituals: Communication in Birds, Mammals and Other Animals*. Cambridge, MA: Harvard University Press.

——— (2003). *Spirit of the Wild Dog*. Sydney: Allen and Unwin.

Rogers, L. J., and Walsh, J. (1982). Short-comings of psychomedical research into sex differences in behaviour: social and political implications. *Sex Roles* 8: 269–281.

Rogers, W. S., and Rogers, R. S. (2001). *The Psychology of Gender and Sexuality*. Buckingham, UK: Open University Press.

Roiphe, K. (1993). *The Morning After: Sex, Fear and Feminism on Campus*. New York: Little, Brown.

Rose, H. (1983). Hand, brain and heart: towards a feminist epistemology for the natural sciences. *Signs: Journal of Women in Culture and Society* 9: 73–96.

——— (1994). *Love, Power and Knowledge*. Cambridge: Polity Press.

——— (2000). Colonizing the social sciences? In *Alas Poor Darwin: Arguments Against Evolutionary Psychology*, ed. H. Rose and S. Rose, pp. 127–153. New York: Harmony Books.

Rose, H., and Rose, S. (1969). *Science and Society*. Harmondsworth, UK: Penguin.

——— (1976). *The Political Economy of Science*. London: Macmillan.

——— (1978). The IQ Myth. *Race and Class* XX(1): 63–74.

———, eds. (2000). *Alas Poor Darwin: Arguments Against Evolutionary Psychology*. New York: Harmony Books.

Rose, S. (1976). Scientific racism and ideology: the IQ racket from Galton to Jensen. In *The Political Economy of Science*, ed. H. Rose and S. Rose, pp. 112–141. London: Macmillan.

——— (1982). From causations to translations: a dialectical solution to a reductionist enigma. In *Towards a Liberatory Biology*, ed. S. Rose, pp. 10–25. London: Allison & Busby.

——— (1984). Biological reductionism: its roots and social functions. In *More Than the Parts: Biology and Politics*, ed. L. Birke and J. Silvertown, pp. 9–32. London: Pluto.

——— (1995). The rise of neurogenetic determinism. *Nature* 373: 380–382.

——— (1997). *Lifelines: Biology, Freedom, Determinism*. London: Penguin.

Ross, H. L. (1972). Odd couples: homosexuals in heterosexual marriages. *Sexual Behaviour* 2(7): 42–49.

Ross, M. W. (1983a). Feminity, masculinity, and sexual orientation: some cross-cultural comparisons. *Journal of Homosexuality* 9: 27–36.

——— (1983b). *The Married Homosexual Man: A Psychological Study*. Melbourne: Routledge and Kegan Paul.

Rossi, A. (1965). Women in science. Why so few? *Science* 148: 1196–1202.

——— (1977). A biosocial perspective on parenting. *Daedalus* 106: 1–32.

Rousseau, J. J. (1762). *Émile, or Education*, Everyman Edition. London: Dent, 1972.

Rowe, W. S. G. (1962). Homosexuality. *Medical Journal of Australia* II (11): 321.

Rowse, A. L. (1983). *Homosexuals in History: A Study of Ambivalence in Society, Literature and the Arts*. London: Dorset.

Rubin, G. (1984). Thinking sex. In *Pleasure and Danger: Exploring Female Sexuality*, ed. C. Vance, pp. 267–319. Boston: Routledge.

Ruse, M. (2001). *The Evolution Wars: A Guide to the Debates*. New Brunswick: Rutgers University Press.

Rushton, J. P. (1994). *Race, Evolution and Behavior*. New Brunswick, NJ: Transaction.

——— (1999). *Race, Evolution and Behavior* (abridged). New Brunswick, NJ: Transaction.

Russell, B. (1945). *A History of Western Philosophy and Its Connection with Political and Social Circumstances from the Earliest Times to the Present Day*. London: Allen and Unwin.

Salinger, L. J. (1970). *The Elizabethan Literary Renaissance: Pelican Guide to English Literature*. Harmondsworth, UK: Penguin.

Salzman, F. (1977). Are sex roles biologically determined? *Science for the People* 4: 27–43.

Sanders, G., and Kadam, A. (2001). Prepubescent children show the adult relationship between dermatoglyphic asymmetry and performance on sexually dimorphic tasks. *Cortex* 37: 91–100.

Sanders, G., and Waters, F. (2001). Fingerprint asymmetry predicts within sex differences in the performance of sexually dimorphic tasks. *Personality and Individual Differences* 31: 1181–1191.

Sandoval, C. (2000). *Methodology of the Oppressed.* Minneapolis, MN: University of Minnesota Press.

Savulescu, J. (2001). Procreative beneficience: why we should select the best children. *Bioethics* 15: 413–426.

Saxonhouse, A. (1976). The philosopher and the female in the political thought of Plato. *Political Theory* 4: 195–212.

——— (1984). Eros and the female in Greek political thought: an interpretation of Plato's *Symposium*. *Political Theory* 12: 5–27.

Sayers, J. (1982). *Biological Politics: Feminist and Anti-Feminist Perspectives*. London and New York: Tavistock.

Schleidt, M. (1992). The semiotic relevance of human olfaction: a biological approach. In *Fragrance: The Psychology and Biology of Perfume*, ed. S. Van Toller and G. H. Todd, pp. 37–50. London: Elsevier.

Schwartz, H. S. (2001). *The Revolt of the Primitive: An Inquiry into the Roots of Political Correctness.* Westport, CT: Praeger.

Schwartz, J. (1984). *The Sexual Politics of Jean-Jacques Rousseau.* Chicago: University of Chicago Press.

Scollard, J. (1983). *No-Nonsense Management Tips for Women.* New York: Pocket Books.

Scott, J. (1988). Deconstructing equality versus difference. *Feminist Studies* 14: 33–50.

Seidman, S. (1995). Deconstructing queer theory or the under-theorization of the social and ethical. In *Social Postmodernism: Beyond Identity Politics*, ed. L. Nicholson and S. Seidman, pp. 116–141. New York: Cambridge University Press.

Shackelford, T. K., and Larsen, R. J. (1999). Facial attractiveness and physical health. *Evolution and Human Behavior* 20: 71–76.

Shanley, M. L. (1981). Marital slavery and friendship: John Stuart Mill's *The Subjection of Women. Political Theory* 9: 229–247.

Shaw, E., and Burns, A. (1993). Guilt and the working parent. *Australian Journal of Marriage and the Family* 14(1): 30–43.

Shaywitz, B., Shaywitz, S. E., Pugh, K. R., et al. (1995). Sex differences in the functional organization of the brain for language. *Nature* 373: 607–609.

Sherwin, B. B. (1988). A comparative analysis of the role of androgen in human male and female sexual behavior: behavioral specificity, critical thresholds, and sensitivity. *Psychobiology* 16: 416–425.

Shute, V. J., Pelligrino, J. W., Hurbert, L., and Reynolds, R. W. (1983). The relationship between androgen levels and human spatial abilities. *Bulletin of the Psychonomic Society* 21: 465–468.

Silliman, J., and King, Y. (1999). *Dangerous Intersections: Feminist Perspectives on Population, Environment, and Development.* Cambridge, MA: South End.

Simpson, M. J. A. (1983). Effect of sex of an infant on the mother–infant relationship and the mother's subsequent reproduction. In *Primate Social Relationships: An Integrated Approach*, ed. R. A. Hinde, pp. 53–57. Oxford, UK: Blackwell.

Slaughter, J., and Kern, R. (1981). *European Women on the Left: Socialism and Feminism, and the Problem Faced by Political Women, 1880 to the Present.* Westport, CT: Greenwood.

Slijper, F. (1984). Androgens and gender role behaviour in girls with congenital andrenal hyperplasia. In *Progress in Brain Research*, ed. G. J. De Vries, J. P. C. DeBruin, H. M. B. Uylings, and M. A. Corner, pp. 417–422. Amsterdam: Elsevier.

Smith, B. (1985). Towards a black feminist criticism. In *The New Feminist Criticism*, ed. E. Showalter, pp. 168–185. New York: Pantheon.

Snitow, A. (1991). Gender diary. In *Conflicts in Feminism*, ed. M. Hirsch and E. F. Keller, pp. 9–43. New York: Routledge.

Sodersten, P. (1976). Lordosis behavior in male, female, and androgenized female rats. *Journal of Endocrinology* 70: 409–420.

Solinger, R., ed. (1998). *Abortion Wars: A Half Century of Struggle, 1950–2000*. Berkeley: University of California Press.

Sommers, C. H. (2000). *The War Against Boys: How Misguided Feminism Is Harming Our Young Men*. New York: Simon & Schuster.

Soper, K. (1990). Feminism, humanism and postmodernism. *Radical Philosophy* 55: 11–17.

Spanidou, I. (1986). *God's Snake*. London: Secker and Warburg.

Spencer, H. (1864–1867). *Principles of Biology*, 2 vols. London: Williams and Norgate.

Spitzka, E. A. (1903). A study of the brain of the late Major J. W. Powell. *American Anthropology* 5: 585–643.

Spivak, G. (1990). *In Other Worlds*. New York: Routledge.

SPLC (1999). *Intelligence Report*. Special issue. *1998. The Year in Hate* 93 (Winter): 30. Southern Poverty Law Center, Montgomery, AL.

Stanton, T., ed. (1884). *The Woman Question in Europe: A Series of Original Essays*. New York: G. P. Putnam's Sons.

Stanton, W. (1960). *The Leopard's Spots: Scientific Attitudes towards Race in America 1815–1859*. Chicago: University of Chicago Press.

Stein, E. (1993). *Forms of Desire: Sexual Orientation and the Social Constructionist Controversy*. London: Routledge.

Steinberg, S. (1995). *Turning Back: The Retreat from Racial Justice in American Thought and Policy*. Boston: Beacon.

Storr, A. (1964). *Sexual Deviation*. Harmondsworth: Penguin.

——— (1970). Human aggression. *Psychiatric Communications* 12: 27–30.

Suleiman, S., ed. (1986). *The Female Body in Western Culture: Contemporary Perspectives*. Cambridge, MA and London: Harvard University Press.

Swaab, D. F., and Hofman, M. A. (1990). An enlarged suprachiasmatic nucleus in homosexual men. *Brain Research* 537: 141–148.

——— (1995). Sexual differentiation of the human hypothalamus in relation to gender and sexual orientation. *Trends in Neurosciences* 18: 264–270.

Swyer, G. I. M. (1968). Clinical effects of agents affecting fertility. In *Endocrinology and Human Behaviour*, ed. R. P. Michael, n.p. Oxford, UK: Oxford University Press.

Taçon, P., and Pardoe, C. (2002). Dogs make us human. *Nature Australia* 27(4): 53–61.

Tanenbaum, L. (1999). *Slut! Growing Up Female with a Bad Reputation*. New York: Seven Stories.

Tatalovich, R. (1997). *The Politics of Abortion in the United States and Canada: A Comparative Study*. New York: M. E. Sharpe.

Taylor, J. (1992). *Paved with Good Intentions. The Failure of Race Relations in Contemporary America*. New York: Carroll and Graf.

Theweleit, K. (1987). *Male Fantasies, Vol. 1: Women, Floods, Bodies, History*, trans. S. Conway, E. Carter and C. Turner. Minneapolis, MN: University of Minnesota Press.

Thompson, D. (1985). *Flaws in the Social Fabric: Homosexuals and Society in Sydney*. Sydney: Allen and Unwin.

Thompson, P. M., Cannon, T. D., Narr, K. L., et al. (2001). Genetic influences on brain structure. *Nature Neuroscience* 4(12): 1253–1258.

Thompson-Handler, N., Malenky, R. K., and Badrian, N. (1984). Sexual behavior of *Pan paniscus* under natural conditions in the Lomako Forest, Equateur, Zaire. In *The Pygmie Chimpanzee*, ed. R. L. Shufman, pp. 347–368. New York: Plenum.

Thornhill, R., and Gangestad, S. W. (2000). Facial attractiveness. *Trends in Cognitive Sciences* 3: 452–460.

Thornhill, R., Gangestad, S. W., and Comer, R. (1995). Human female orgasm and mate fluctuating asymmetry. *Animal Behaviour* 50: 1601–1615.

Thornhill, R., and Palmer, C. (2000). *A Natural History of Rape: Biological Bases of Sexual Coercion*. Cambridge, MA: MIT Press.

Tiger, L., and Fox, R. (1972). *The Imperial Animal*. London: Secker and Warburg.

——— (1978). The human biogram. In *The Sociobiology Debate*, ed. A. L. Caplan, pp. 57–63. New York: Harper & Row.

Tooby, J., and Cosmides, L. (1990). The past explains the present: emotional adaptations and the structure of ancestral environments. *Ethology and Sociobiology* 11: 375–424.

——— (1992). The psychological foundations of culture. In *The Adapted Mind: Evolutionary Psychology and the Generation of Culture*, ed. J. Barkow, L. Cosmides, and J. Tooby, pp. 19–136. New York: Oxford University Press.

Trevor-Roper, H. R. (1960). *The European Witch-Craze of the 16th and 17th Centuries*. Harmondsworth, UK: Penguin.

Trinh, T. (1989). *Woman, Native, Other*. Bloomington: Indiana University Press.

Trivers, R. L. (1972). Parental investment and sexual selection. In *Sexual Selection and the Descent of Man: 1871–1971*, ed. B. Campbell, pp. 136–179. Chicago: Aldine.

——— (1978). The evolution of reciprocal altruism. In *The Sociobiology Debate*, ed. A. L. Caplan, pp. 213–226. New York: Harper & Row.

Trivers, R. L., and Willard, D. (1973). Natural selection of parental ability to vary the sex ratio of offspring. *Science* 179: 90–92.

Truell, P. (1996). Success and sharp elbows: one woman's path to lofty heights on Wall Street. *New York Times*, July 2: D1/D4.

Urbinati, N. (1991). John Stuart Mill on androgyny and ideal marriage. *Political Theory* 19: 626–648.

Valian, V. (1999). *Why So Slow? The Advancement of Women*. Cambridge, MA: MIT Press.

Vines, G. (1992). Obscure origins of desire. *New Scientist*, November 22, pp. 2–8.

——— (1993). *Raging Hormones: Do They Rule Our Lives?* Berkeley: University of California Press.

vom Saal, F. S. (1983). The interaction of circulating oestrogens and androgens in regulating mammalian sexual differentiation. In *Hormones and Behaviour in Higher Vertebrates*, ed. J. Balthazart et al. Berlin and New York: Springer Verlag.

Wade, I. O. (1941). *Studies on Voltaire*. Princeton: Princeton University Press.

Walker, Alice (1983). *In Search of Our Mothers' Gardens: Womanist Prose*. Orlando, FL: Harcourt Brace Jovanovich.

Wallen, K. (1990). Desire and ability: hormones and the regulation of female sexual behavior. *Neuroscience and Biobehavioral Reviews* 14: 233–241.

Walters, W. A. W., and Ross, M. W., eds. (1986). *Transsexualism and Sex Reassignment*. Melbourne: Oxford University Press.

Walton, P. (1982). The culture in our blood. *Women: A Journal of Liberation* 8(1): 43–45.

Walzer, M. (1974). *The Revolution of the Saints: A Study on the Origins of Radical Politics*. New York: Atheneum.

Wandor, M. (1981). *Understudies, Theatre and Sexual Politics*. London: Eyre Methuen.

Ward, I. L. (1992). Sexual behavior: the product of perinatal hormonal and prepubertal social factors. In *Sexual Differentiation: Handbook of Behavioral Neurobiology*, vol. 11, ed. A. A. Gerall, H. Moltz, and I. L. Ward, pp. 157–180. New York: Plenum.

Ward, I. L., and Weisz, J. (1980). Maternal stress alters plasma testosterone in fetal males. *Science* 207: 328–329.

Warrack, J. (1973). *Tchaikovsky*. London: Hamisch Hamilton.

Wasser, L. M., and Wasser, S. K. (1995). Environmental variation and developmental rate among free ranging yellow baboons (*Papio cynocephalus*). *American Journal of Primatology* 35: 15–30.

Watson, L. (1979). Homosexuals. In *Mental Disorder and Madness*, ed. E. M. Bates and P. R. Wilson, pp. 134–161. St. Lucia, Queensland: University of Queensland Press.

Weber, M. (1930). *The Protestant Ethic and the Spirit of Capitalism*, trans. T. Parsons. New York: Charles Scribner's Sons.

Wedekind, C., and Furi, S. (1997). Body odour preferences in men and women—do they aim for specific MHN combinations or simply heterozygosity? *Proceedings of the Royal Society of London—Series B: Biological Sciences* 264: 1471–1479.

Wedekind, C., Seebeck, T., Bettens, F., and Paepke, A. J. (1995). MHC dependent mate preferences in humans. *Proceedings of the Royal Society of London—Series B: Biological Sciences* 260: 245–249.

Weeks, J. (1981). *Sex, Politics and Society: The Regulation of Sexuality Since 1800*. London and New York: Longman.

Weinberg, T. S. (1984). Biology, ideology, and the reification of developmental stages in the study of homosexual identities. *Journal of Homosexuality* 10: 77–84.

Weiner, T. S., and Wilson-Mitchell, J. E. (1990). Nonconventional family life-styles and sex-typing in six-year-olds. *Child Development* 61: 1915–1934.

Weiss, P. (1987). Rousseau, antifeminism and woman's nature. *Political Theory* 15: 81–98.

——— (1990). Sex, freedom and equality in Rousseau's *Émile*. *Polity* 22: 603–625.

Wells, L. (1987). *Know How: 36 Australian Women Reveal Their Career Success Secrets*. Sydney: Australasian Publishing Company.

West, D. J. (1977). *Homosexuality Re-Examined*. London: Duckworth.

Whitford, M. (1991). *Irigaray: Philosophy in the Feminine*. London: Routledge.

Wickelgren, I. (1999). Genetics—discovery of "gay gene" questioned. *Science* 284: 571.

Will, J. A., Self, P. A., and Datan, N. (1976). Maternal behavior and perceived sex of infant. *American Journal of Orthopsychiatry* 46: 135–139.

Williams, T. J., Pepitone, M. E., Christensen, S. E., et al. (2000). Finger length ratios and sexual orientation. *Nature* 404: 455–456.

Wilson, E. O. (1975). *Sociobiology: A New Synthesis*. Cambridge, MA: Harvard University Press.

——— (1998). *Consilience: The Unity of Knowledge*. New York: Knopf.

Winant, H. (1997). Behind blue eyes: whiteness and contemporary U.S. racial politics. In *Off White: Readings on Race, Power, and Society*, ed. M. Fine, L. Weis, L. C. Powell, and L. M. Wong, pp. 40–53. New York: Routledge.

Winks, R. W., ed. (1972). *Slavery: A Comparative Perspective—Readings on Slavery from Ancient Times to the Present*. New York: New York University Press.

Winthrop, D. (1986). Tocqueville's American woman and "the true conception of democratic progress." *Political Theory* 14(2): 239–261.

Witelson, S. F. (1989). Hand and sex differences in the isthmus and genu of the human corpus callosum. *Brain* 112: 799–835.

Wittig, M. (1989). On the social contract. *Feminist Issues* 9: 3–12.

Wolf, N. (1991). *The Beauty Myth: How Images of Beauty Are Used Against Women*. New York: William Morrow.

Wollstonecraft, M. (1975). *Vindication of the Rights of Woman*, ed. M. Kramnick, based on the 1792 edition. Harmondsworth, UK: Penguin.

Women, Science and Society (1978). Special issue published by *Signs: Journal of Women in Culture and Society*, issue 4.

Woolf, V. (1931). *Professions for Women.* In *Virginia Woolf: Women and Writing*, ed. M. Barrett. London: Women's Press, 1979.

Wurtzel, E. (1997). *Prozac Nation: Young and Depressed in America.* New York: Riverhead Books.

Young, E. W. (1988). Nurses' attitudes towards homosexuality: analysis of change in AIDS workshop. *Journal of Continuing Education in Nursing* 19: 9–12.

Young, I. M. (1986). Impartiality and the civic public: some implications of feminist critiques of moral and political theory. *Praxis International* (special issue, entitled *Feminism as Critique*, ed. S. Benhabib and D. Cornell). 5: 382–389.

Zahavi, A., and Zahavi, A. (1997). *The Handicap Principle.* Oxford, UK: Oxford University Press.

Zhang, S., and Odenwald, W. F. (1995). Misexpression of the white (w) gene triggers male-male courtship in drosophila. *Proceedings of the National Academy of Sciences* 92: 5525–5529.

Index

abortion, 175
after rape, 91–92
government intervention in, 41, 114
as prenatal selection, 16, 18–19, 98–99, 212, 222
Adiós Barbie (Edut), 106
Adorno, T., 99
age, 30–31, 48, 76. *See also* development
agency, 91, 92, 94
aggression, 94, 153, 165
evolution of, 98–99, 127–133
as genetically caused male trait, 14, 37, 66, 92–93
sociobiology on, 89–90
vagueness of definition of, 90–91
Agreement of the People, 94–95
AIDS/HIV, 195, 199–200, 201–203
Alas Poor Darwin (Rose and Rose), 232
alcoholism, genetic basis for claimed, 3, 14, 194
Allen, L., 77
Allen, S., 202
altruism, gene for, 208
Anderson, B., 206
androgyny, 185
anogenital licking, 53–56, 76
antisocial behavior, genetic basis for claimed, 9, 16, 38, 194
anxiety, effects on test scores, 134–135
Arendt, H., 87, 95, 103, 229
Aristotle, 91, 231
Attis, myth of, 142

Bacchus, 185
Bacon, F., 8
Bailey, J. M., 210–211, 217–219
Baker, R. R., 181
Baker, S., 150
Balaban, E., 216
Barash, D., 36
Bardwick, J. M., 111
Beach, F. A., 78, 146
Beauty Myth, The (Wolf), 106
behavior, 19, 37, 62, 173, 203. *See also* brain function; sexual behavior
  biological basis for, 35–36
  in chain of influence with genes, 12, 48, 74
  complexity of, 13–14, 64–65, 230
  desire to explain by genes, 5, 12
  effects of hormones on, 48, 74, 111, 150–151
  evolutionary psychology on, 5–7
  expectation of manipulation of, 14–16
  freedom to choose, 22–24, 30, 91, 226–227
  and gene-mapping, 11–12
  genes as cause of, 3–4, 66, 111, 194, 208–209, 226–227
  levels of research into and explanations for, 20–22
  parental behavior shaping sex differences in, 58–59, 153–154
  predispositions/susceptibilities toward, 15–16, 208–209
  relations among disciplines in explaining, 20–21
  search for genetic basis of, 14–15, 17
  sex differences in, 42, 48–49, 67, 110–111, 148
behavioral genetics. *See* evolutionary psychology
Bellis, M. A., 181
Benbow, C., 134–136
Benhabib, S., 87
Benjamin, J., 160
Benke, T., 137
Bereczkei, T., 183
Berenbaum, S., 150–151
Bioethics (Savulescu), 18
biological determinism. *See* determinism
biological essentialism, 100. *See also* essentialism
biology, 63, 94, 236. *See also* science; sociobiology
  determinism in, 31–32, 36–39, 42, 100
  feminists on, 43, 95, 100
  oppressive uses of, 36, 109–110, 116, 232–234
  reductionism in, 29–30, 64–65
  relation to society, 25, 36, 97–98, 156–157
  of sex differences, 118–119, 149
  in sexual orientation, 212–213
  vs. adaptability, 99, 227–228
Birke, L., 104
Bishop, K. M., 77–78
Bloch, E., 130, 169
Bloom, H., 132
Blum, D., 229, 231
body, 87, 198
  attractiveness of, 158–160, 178–183
  development of sex differences in, 131–132

ornaments to enhance mating, 173, 178
sex differences in size, 174–177
sex similarities in physical stamina, 131–132
sexualization of, 157–159
studies of fingers and fingerprints, 220–223
symmetry of, 180–182
variability in, 117–118
body/mind dichotomy, 8–9, 88
Boster, J. S., 177
Bouchard, T., 123
Bourdieu, P., 214–215
Bower, B., 216
Braidotti, R., 104
brain, 12, 146, 196, 228. *See also* mind; Stone-Age brain
effects of gay gene in, 210–211
hormones' effects on, 49–53, 133, 170–172, 199–200, 203
labeling of inferiorities in, 32–35, 221
plasticity of, 5, 76
sex differences in, 7, 61, 100, 148
brain development, 73, 146
effects of sex hormones on, 52, 78, 133, 202
sensitive periods of, 149–151
brain function, 48, 128, 222. *See also* behavior; math abilities; spatial abilities; verbal abilities
control of sex hormones as, 73–74
flexibility in, 70–71, 73
genetic module theory of, 67–70
hemispheric specialization, 72–73
imaging of, 19–20
influences on, 53–54, 62, 68
sex differences in, 19–20, 50–51, 77–78, 117, 134–138
and sexual orientation, 201–203
brain size
intelligence's relation to, 33–34
sex differences in, 61, 77–78, 109–110
brain structure, 48, 73
corpus callosum, 74–76
and homosexuality, 201–203, 203
influences on, 53–54, 62
and partner choice, 20, 201–202
sex differences in, 35, 50–51
Breedlove, S. M., 71
Broca, P., 33–34
Brown, G. R., 37
Buffery, A. W. H., 111
Burstyn, V., 103
Burt, C., 6, 217
Bush, G., 41
Bush, G. W., 41
Buss, D., 69

Campbell, A., 58–60, 161
canalization, of genes, 69
Caplan, P., 125
Caravaggio, M. da, 185
castrati, 186–187
castration
chemical, 40, 196
effects of, 51–52, 54
Cattell, R., 18
causality, 12, 29
debate over direction of, 234–235
in reductionism, 29–30
unfounded assumptions of, 127, 137
children, different treatment of, 57–59

Chorover, S. L., 36
Cicero, M. T., 96
Cixous, H., 44
Clarke, L., 110
class, social, 36, 41, 60, 101–102
clothing, 57, 185–189
codes, gene, 47–48
cognitive processing. *See* brain function
competition, 174–175. *See also* aggression
Comte, A., 195
Consilience: The Unity of Knowledge (Wilson), 12
constructionism, 43, 153–154
contraception, 41, 114, 143
contract theory, gender relations as, 87–88, 102
Cooley, C. H., 24
Coonan, H., 205
Copeland, P., 194, 230
Cornell, D., 103
cortisol, 126. *See also* hormones
Cosmides, L., 69, 89–90, 127
Crittenden, D., 106
écriture feminine movement, 43–44
culturalism, 231–234
culturgens, 89
"Cyborgs" (Haraway), 107

Darwin, C., 32–33
Dawkins, R., 36–37, 172
de Lacoste, M.-C., 77
de Lacoste-Utamsing, C., 77
De Legibus (Cicero), 96
Declaration of Independence, 95
Declaration of Rights for Women, 95
Declaration of Sentiments, 95
Delgado, J. M. R., 196
Denenberg, V., 75
depression, 3, 21–22
Derrida, J., 99
Descartes, R., 88
determinism, 27, 29–30, 42, 77, 153, 233. *See also* gene expression
  biological, 4–5, 31–33, 36–39
  of feminists, 100–101
  genetic, 207
  in recapitulation theory, 35–36
  vs. free will, 22–24
development, 125. *See also* brain development; sex differences
  arrested, 32–35, 199
  causality in, 234–235
  effects of hormones in, 52, 55–56, 118
  effects of sex-differentiated treatment on, 56–58
  experience in, 77, 227–228
  flexibility of, 6
  of gender identity, 153, 161
  interaction among genes, hormones, and experience in, 5, 75–76, 78
  of nerve cells, 51, 71
  sensitive periods of, 149–152
development neuroscience, 5
dichotomies, 31–33, 104. *See also* sexual dimorphism
Dickens, H., 187
Dickinson, R. L., 198
diseases/disorders, 64
  genes claimed as cause of, 3, 193
  homosexuality treated as, 198–199, 203–204
  model of, 13–14, 203–204

reductionism in, 21–22
sex differences in, 158–161
DNA, 11–12, 47, 65. *See also* genes
Dörner, G., 52–53, 151–152, 199, 220–221
Duchaine, B., 68
dyslexia, gene for claimed, 3

Émile (Rousseau), 84–86
economics, rationalism in, 90
education
role of IQ test scores in, 120–121
sex differences in school performance, 123–124
women's, 109–110, 117, 133–134
Edut, O., 106
egalitarianism, not dependent on sameness, 96–98, 104
egg (ova), 91
genetic mutations from, 143–144
relative importance attached to, 141, 143
Ehrhardt, A. A., 118, 150, 198–199
Eisenstein, Z., 104
Ellis, H., 117
emotions, 138, 163, 166
gendered allocation of, 88–89, 92–93
and women as nurturers, 111–112
environment, 12, 16, 63, 68, 144, 153, 223. *See also* experience
and direction of causality, 234–235
and expression of genes, 6–7, 65, 209
and hormones, 50, 73–74
individual's selection of, 13, 60–61, 123–124
interaction with genes and hormones, 48, 78, 148–149
and sexual orientation, 212–213, 218–219
weight of influence attributed to, 13, 30, 60–61, 228
epigenetic constraints, 6, 12
equality, 197. *See also* oppression
with difference, 43, 100–101, 111
effects of women's movements on, 105–106
futility of, 38, 40
in gender relations, 86–89, 95–96
for homosexuals, 206–207
illegality of gender distinctions, 103–104
and inequality, 84, 119
Eskimos, spatial abilities of, 125
essentialism, 27–29
biological, 100
in feminism, 43, 102–103, 106–107
of gender, race, and age, 30–31
on sex differences, 33–34
estrogen, 49–50. *See also* hormones
eugenics, 16, 18, 36, 41, 114, 197
Evans, D. T., 197
evolution, 28, 132, 165, 208
in explanation of sex and race differences, 7, 32–34, 40, 109, 232–233
and human adaptability, 227–228
in recapitulation theory, 35
sex differences in genetic mutations, 143–144
of sexual dimorphism, 176–177
evolutionary biology, 6
evolutionary psychology, 4–8, 43, 83, 90, 92, 107, 173
appeal of, 10, 107
essentialism in, 28–29

evolutionary psychology (*continued*)
on evolution of the mind, 40, 127–133
gendered allocation of emotions by, 92–94
on genes' control of behavior, 11–12, 58, 66, 111
on genetic control of the mind, 67–69
on human adaptability, 227–228
on influences on mate choice, 179–180, 183
role among sciences, 231–232
on role of environment, 60–61
on sex differences, 67, 125
as supremacist, 39, 229–230
experience, 64, 123. *See also* environment
in development of sex differences, 53, 55–56, 61–62, 74–76
and direction of causality, 234–235
effects on brain, 73–74, 202
influence of ignored, 30, 53, 77
interaction with genes and hormones, 5, 48, 70–76
role of, 8, 227–228
Eysenck, H., 6, 41, 121

family, 14
complexity of relationships in, 154–155
in development of gender identities, 152–154
sex differences in treatment of children in, 57–59
in studies on homosexuality, 209–210, 217–220
Fausto-Sterling, A., 216
Feingold, A., 123–125
femininity. *See* masculinity/femininity
feminist criticism, 86–87
feminism
determinism in, 5, 38–39, 100–101
divisions within, 101–102
essentialism in, 102–103, 106–107
goals of, 98, 225–226
in history, 95–96
on meanings of "woman" and "gender," 99, 104
on sex *vs.* gender, 43–45, 102
and sociobiology, 8, 42–43
and women's movements, 88, 94, 104–107, 111
and women's sexuality, 113–116
fertility, 141–142, 198. *See also* reproduction
flexibility, 25, 28, 149
denied by determinism, 5, 24–25, 30
of nervous system, 5, 62, 70–71
Flynn, J. R., 121
Foucault, M., 84, 87, 99, 161
Fox, R., 40, 111
Fraser, N., 99
freedom, 22–24, 95, 98, 236
French Revolution, 94–95

Gagnon, P., 190
Galaburda, A. M., 118, 133, 199
Gallup, G. G., Jr., 177
Galton, F., 35
Gatens, M., 44
Gaulin, S. J. C., 177
"gay gene," 8, 201–203. *See also* homosexuality
flaws in studies of, 213–216

implications of, 206–209, 222–223
lack of evidence of, 223–224
search for, 209–210, 219, 230
gender, 11, 83
double-play with, 186–189
and essentialism, 30–31, 102–103
homosexuals as outside norms of, 200–201
meanings and use of, 43–45, 99
oppression based on, 103–104
symbolisms of, 156–157
gender identity, 195
and complexities in determining sex, 145–147
development of, 152–154, 161
gender relations, 93. *See also* mating
as contract theory, 87–88
effects of assumptions of sex differences in, 30, 66–67, 98–99
in history, 84–86, 94–95, 114
in love and marriage, 166–168
power in, 84–86
sexuality in, 84, 113–114, 116
gender roles, 43, 88, 147
development of, 40, 127–133, 153
women as nurturers, 111–112
women refuting sociobiology on, 42–43
gender schemas, 231
gender/sex split, 102–104, 160
gene expression, 12, 17, 48, 65, 208
environment and, 6–7, 60–61, 63, 70–73
as essential, 23–24
technologies to measure, 19
vs. free will, 22–24
genes, 70, 132, 193. *See also* DNA
aggression linked to, 89–90, 127–128, 130
for altruism, 208
causing only predisposition for behavior, 15–16, 208–209
in chain of influence with behavior, 3–4, 12, 48, 74, 194, 234–235
and choice of environments, 123–124, 153
and cognition, 60–61, 67–69, 121–122
and complexities in determining sex, 145–146
complexity of effects of, 11, 230
in development of sex differences, 74–76, 147–148
drive to pass on, 39–40, 90–92, 165, 172–173
in essentialism, 28–29
fitness of, 15–16, 97–99, 180–181
funding for research on effects, 64–65, 194
inadequacy to explain behavior, 14, 21, 29–30
interweaving of influences on behavior, 5, 74–76, 78, 111, 148–149
and loss of power to choose, 226–227
mapping technology, 11–12
and mate choice, 179, 183–184
modules of, 67–70
number of, 65
search for basis of behaviors in, 12, 14–15, 17
sex differences explained by, 3–5, 20–21, 29–30, 37, 42, 107

genetic mutations, sex differences in, 143–144
genetic susceptibilities, 15–16
genetics
  attempts to manipulate, 15–16, 35–36
  equated with inevitability, 99
  use of twin studies in, 216–220
genitalia, 57, 175, 177
  and complexities in determining sex, 145–147
  development of differences in, 147–152
  differences in size, 176, 198
  influences on, 49–55, 150–151
genome mapping attempts, 14–15, 17
Geschwind, N., 118, 133, 199
God's Snake (Spanidou), 85–86
Goethe, J. W. von, 187
Gorski, R. A., 202
Gray, J. A., 111
Guille-Escuret, G., 10
Gurr, R. C., 111

Habermas, J., 236
Haier, R., 134–136
hair, 157–158
Hamer, D. H., 194, 209–213, 215–216, 230
Haraway, D. J., 44, 107, 226
Haug, F., 157–159
Hegel, G. W. F., 89
Heller, A., 43, 90
hemispheric specialization, 72–73, 200. *See also* lateralization
  and corpus callosum, 75–76
  sex differences in, 134–137
Herrnstein, R. J., 41
Hess, R., 35
heterosexuality, 88, 113–114
Hines, M., 150–151, 231
history
  feminism in, 95–96
  homosexuality in, 195–196
  of oppression of homosexuals, 204–207
  periods of relative gender equality in, 94–95
  racism and sexism in, 32
  structure of masculinity in, 86–89, 93–94
  of study of sex differences, 230–231
History of Sexuality, A (Foucault), 84
Hoagland, S., 39
Hobbes, T., 87, 89, 92
Hofman, M. A., 203
Holloway, R. L., 77
homelessness
  gene for claimed, 3, 38
  women's, 86, 94
homosexuality, 11, 167. *See also* "gay gene"
  attitudes toward, 143, 205–207, 212
  genetic basis for claimed, 3, 14, 194, 201–202
  in history, 195–196
  hormones claimed to cause, 52, 78, 150–152
  male *vs.* lesbianism, 205
  medical model not applicable to, 194–195
  oppression based on, 102, 204–207
  research on, 197–203, 213–220
  and studies of fingers and fingerprints, 220–223
  treatment for, 196, 203–204

Homosexuality Re-Examined (West), 198
Homosexuality (West), 198
Honig, B., 103
hormones
  and complexities in determining sex, 145–146
  in development of sex differences, 55–56, 74–76, 147–151
  in development of spatial abilities, 125–126
  in differences in fingers and fingerprints, 220–223
  effects of, 21–22, 49–53, 73, 110–111, 179, 230
  effects on brain, 61, 203
  fluctuations in levels of, 73–74, 151–152
  and homosexuality, 196, 198–203
  influence on sexual behavior, 169–172
  influences on, 54–55
  interaction with genes and environment, 71, 73–76, 78, 111, 234–235
  interactions with behavior, 55, 62
  and menstrual cycles, 21–22, 110–111
  receptors for, 49–50
  sex differences attributed to, 29–30, 42, 48
Hoult, T. F., 184
How the Mind Works (Pinker), 66
Hubbard, R., 224
Human Genome Project, 17, 64, 65
Human Genome Sequence Consortium, 143
human life, meaning of, 91, 236
hunting, 40, 90, 125, 127–133
Huschke, E., 33
Hutt, C., 111, 117–118
Hyde, J., 124

Illich, I., 195
imprinting, 59
information processing. *See* brain function
instinct, *vs.* learning, 227
intellect, 23, 91, 109–110, 119–120
intelligence, 24, 38, 120
  attributed to men, 37, 118
  genetic basis for claimed, 6, 14, 194
  relation to brain size, 33–34
  research on, 117, 217
IQ test scores. *See also* intelligence
  sex differences in, 120–123
  usefulness of, 120–122
  variability in, 117–118
Irigaray, L., 44

Jeffreys, S., 102
Jensen, A., 6
Jessel, D., 127–128, 131
Jews, 114

Kamin, L., 217
Kaplan, J., 6
Kennedy, R., 205
Kertesz, A., 137
Kimura, D., 96–98, 119, 125, 137
Kinsbourne, M., 72
Kinsey, A., 197–198, 213
Kinsey Report, 116
Kowner, R., 178–179, 182
Kurth, G., 227

La Vigne, A., 10
labor
  differences in, 37, 87, 154
  effects of women's movements on, 105–106
  hunting's role in development of roles in, 127–133
  stamina for, 131–132
  women in public sphere, 110–114, 116, 124
Lai, C. S. L., 70
Lambert, H. H., 100
language. *See also* verbal abilities
  brain regions used for, 136–137
  essentialism in, 30–31
  genetic basis for claimed, 70, 194
  "sex" *vs.* "gender" in, 43–44
lateralization. *See also* hemispheric specialization
  and cause of homosexuality, 199–200
  sex differences in, 77–78
Leacock, E., 130
learning, 59, 64, 133, 149. *See also* environment; experience
  effects of, 68, 71–72, 148
  and human adaptability, 227–228
  vs. genes, 60–61, 70
LeBon, G., 33–34
legislation, and reason, 91
Leigh, A., 34
Lesbian Heresy, The (Jeffreys), 102
lesbianism. *See* homosexuality
LeVay, S., 201–202
Levellers, 94–95
Levy, J., 111
Lewontin, R., 12
Li, W.-H., 143
Locke, J., 27–28, 87, 92
Lumsden, C. J., 89
Lytton, H., 58–59

Machiavelli, 87, 89
Makova, K., 143
Males and Females (Hutt), 117–118
Mall, F. P., 117
Mandler, G., 129
"Manifesto for Cyborgs, A" (Haraway), 226
Marcuse, H., 94, 225–226
Marx, K., 87
masculinity/femininity, 104, 156, 163
  in history of ideas, 87–89
  traits mixed within individuals, 184–185, 189
math abilities, 110, 135
  disincentives for girls', 133–134
  on IQ tests, 122–124
mating. *See also* gender relations; sexual attraction
  partner choice, 201–202, 210
  sociobiology on patterns of, 39–40
  strategies, 172–180, 183–184
McCormick, C., 200
McGue, M., 123
memory. *See* brain function; learning
menstrual cycles, 21–22, 73, 91, 110–111
metabolism, inferiority of, 110
Mezey, S. G., 103
Mill, J. S., 87, 89, 93
Miller, E. M., 18
mind, 66. *See also* brain
  evolution of, 127–133
  genetic module theory of, 67–70, 228

modules, of the mind, 67–70, 228
Moir, A., 127–128, 131
molecular genetics, 5, 10–11, 230
  funding for research in, 64–65, 228
  and genome mapping attempts, 15, 17
  new technologies in, 18–19
Money, J., 52–53, 60, 118, 150, 198–199, 221
monogamy/polygamy, sex differences in, 66
Moore, C. L., 54–55, 62, 74
morality
  under egalitarianism, 96–98
  and good wars, 89, 91
  of prenatal selection, 18
  of women, 100–101, 112
  women's movement accused of devaluing, 114–115
Morris, D., 177
Morris, J., 189
Mother–child relationships, 37, 56. *See also* parental behavior
Mouffe, C., 103–104
Mueller,, 93
Murray, C., 41
Myth of the Twentieth Century (Rosenberg), 115

natural selection, 36, 68, 98, 119, 165. *See also* evolution
  in evolutionary psychology, 6, 8
nature/nurture dichotomy, 74, 153
  declared over, 17–18
  interweaving of influences of, 5, 148–149, 226
  lack of usefulness of, 8–9
Nazis, 35–36, 114–115, 195
nerve cells, 51, 54–55, 70–71
Nerve Growth Factor (NGF) hormone, 71
nervous system, 5, 62. *See also* brain
Neumann, E., 156–157
neurobiology, 5
New Right, 38
Newton, I., 23
Nordeen, E. J., 50
Nuffield Council on Bioethics, 194
nurturing. *See* gender roles

odors, 54, 179–180
Okin, S. M., 86
oppression
  based on differences, 31–35, 103–104, 227
  biology used to justify, 25, 100
  of homosexuals, 204–207
  justifications for, 7, 27, 37–38
  multiple bases for, 101–102
  population control as, 41–42
  and sexuality, 112, 116
organizational effect of testosterone, 78
Oyama, S., 5, 99

Palmer, C., 39
Palmer, J. A., 6, 227–228
Palmer, L. K., 6, 227–228
parental behavior, 152, 174
  and development of sex differences, 53–60, 74, 76, 150–151, 153–154
  different toward males and females, 56–59
  effects of, 78, 150–151
Pateman, C., 87–89
patriarchy, as "natural," 38

Pawlowski, B., 179
Pearson, K., 35
Peritore, N. P., 232–236
personality, 157–159, 178, 194
Phoenix, C. H., 78
Pigliucci, M., 6
Pillard, P. C., 210–211, 217
Pinker, S., 66–67, 70, 89–90, 127
plasticity. *See* flexibility
Plato, 87, 231
play, types of, 150–151
Plummer, D., 204–205
politics, 105
  in control of reproduction, 41–42
  of feminism, 101–103
  of gay gene, 207–208
  of sex differences, 37–38, 106–107
Politics of the Gender Gap, The (Mueller), 93
Pollitt, K., 106
population control, 41–42. *See also* abortion; reproduction
positivism, 13, 154
postmodernism, 15, 206–207
power, 9, 93, 99, 154, 236
  in gender relations, 84–87
  not modeled on biology, 97–98
  and the passing on of genes, 90–92
premenstrual syndrome, 110–111
prenatal diagnoses, 16, 18
Professions for Women (Woolf), 225
Protestant Ethic and the Spirit of Capitalism, The (Weber), 38
Pruner, F., 33
psychology, 20–21
psychosurgery, to "cure" homosexuality, 196
race, 36, 41, 110
  and differences in intelligence, 6, 38, 121–122
  essentialism associated with, 30–31
  justifications for discrimination among, 8, 38, 232–233
  oppression based on, 7, 32–35, 101–102, 227–229
Race, Evolution, and Behavior (Rushton), 41
rank. *See* class, social
rape, 39–40, 66, 90–92
rationalism, 90
reason, 91–92
  and egalitarianism, 96–97
  vs. emotions, 88, 109–110
recapitulation theory, 35
reductionism, 9–11, 21–22, 29, 56, 107, 207, 235
  in biology, 31–33, 64–65
  on cause of homosexuality, 198–200, 210
Reinisch, J., 148–149
relationships, 171. *See also* gender relations; mating
reproduction, 8, 34, 106, 110, 143, 166, 197. *See also* population control
  and appearance, 159–160, 178–184
  drive for, 90–92, 180, 235–236
  focus on motherhood, 113–115, 141
  and love and marriage, 167–168
  and mating strategies, 39–40, 128, 172–184
  and sexuality, 164, 181
responsibility, *vs.* determinism, 30

Rhodes, D. L., 104
Rhodes, G., 190
Rice, G., 211
Rich, A., 87
Ridley, M., 90
rights. *See* equality
Rogers, L., 72–73, 234
Roiphe, K., 106
Romney, D. M., 58–59
Rose, H., 25, 232
Rose, S., 13, 21, 64, 196, 232
Rosenberg, A., 115
Rossi, A., 100
Rousseau, J. J., 84–86, 87, 89, 95
Rushton, J. P., 41, 228–229

Savulescu, J., 18
schizophrenia, 3, 14
science, 29, 107, 138. *See also* biology
  errors in, 9–10, 213–220
  and essentialism, 27–28
  marks of honesty in, 223, 231
  motives for research, 11, 14–15, 64–65
  relations among disciplines, 20–21
  research as value-laden, 194–196, 230
  research on homosexuality, 197–203
  selective use of evidence in, 74, 119–120
  society's influence on, 8, 12–13, 32
  uses of findings of, 20, 223–224, 229–230, 232–234
Seidman, S., 206
"selfish gene," 165, 172–173
separatism, 101
serial analysis of gene expression (SAGE), 71
sex, 227
  assignment of, 152, 169, 190
  complexities in determining, 145–147
  differences in parental behavior based on, 52–60
  relative importance of, 141–142, 144
  vs. gender, 43–45
sex differences, 8–10
  attributed to genes and hormones, 3–5, 14, 24, 37–40, 67–68, 111
  in behavior, 48–49, 67, 153–154, 203
  belief in, 36, 96
  biological, 96, 118–119
  in body size, 174–177
  in body-related disorders, 158–161
  in brain development, 61, 72–73
  in brain functions, 19–20, 117, 134–138
  in brain size, 34, 77–78
  causes of, 20–21, 29–30, 60, 78, 203
  in children's play, 57–58
  in chromosomes, 48
  in cognitive abilities, 119–127, 133
  determinism on, 42, 100
  development of, 52–56, 61–62, 74–76, 147–152
  in emotions *vs.* reason, 88–89, 92–93, 163
  essentialism on, 106–107
  evolution of, 109, 131–132

sex differences (*continued*)
  exaggerated polarities in, 72, 144–147
  feminism on, 43, 100, 106–107
  in fingers and fingerprints, 220–222
  in genetic mutations, 143–144
  history of study of, 4–5, 7, 230–231
  and hormones, 49–53, 74
  and parental behavior, 53–60, 74, 76, 152–154
  results of, 32–35, 153–154, 227, 233–234
  in sexuality, 66–67, 163
  symbolisms of, 156–157
  vs. similarities, 72, 138
sex hormones. *See* hormones; testosterone
Sex in Education (Clarke), 110
Sex on the Brain (Blum), 229
sex organs. *See* genitalia
sex roles, 24, 43, 95. *See also* gender roles
sex/gender split, 102–104, 189–190
sexism, 8, 38, 232–234
sexual attraction. *See also* mating
  brain structure and, 201–202
  complexity of, 190–191
  factors in, 183–184
sexual behavior, 146, 201
  effects of, 50, 71
  hormones' influence on, 54–55, 169–172
  influences on, 51–52, 223
  sex differences in, 66–67
sexual dimorphism, 8, 173–176, 190
  causes of, 176–177
  homosexuals in, 200–201
  inflated polarities in, 144–147, 160
Sexual Hygiene (von Gruber), 115
sexual identity, 57, 60. *See also* gender identity
sexual preference/orientation. *See* heterosexuality; homosexuality
sexuality, 84, 106, 117, 161, 163–164, 166
  and mating success, 178–184
  and natural selection, 98–99
  society's influence on, 155, 168–169
  and violence, 90–92
  women's, 109, 112–116, 169
sexualization, of female body, 157–159
Shaywitz, B., 136
Shute, V., 125–126
Signs (journal), 107
slavery, 83–84, 87
Slijper, F., 151
social class. *See* class, social
social control, and biological determinism, 38
social psychology, 33
social roles, 24
social status, 56. *See also* class, social
society, 18, 101–102
  and biology, 25, 36, 97–98
  competition *vs.* cooperation in, 165–166
  in creation of sex differences, 133–134, 153
  cultural influence on cognitive abilities, 121–125
  culturalism, 231–234
  and evolution of sexual dimorphism, 176–177

homosexuality in, 195–197, 204–207, 211–214, 216, 222–223
hunter-gatherer, 127–133
importance of gender in, 42–43, 83, 144, 152, 157
love and marriage in, 167–168
norms of, 155, 225–226
reproduction success *vs.* other achievements in, 235–236
role in sex differences, 8, 24
and science, 12–13, 32, 64, 107
sex differences in, 20, 37, 148–149
shaping sexuality, 168–169
structure of masculinity in, 86–89
use of science findings, 90–91, 194
sociobiology, 4, 6–8, 36–40, 83, 228
on aggression, 89–90, 94
on genes' and hormones' influence, 111, 179
on hunting in evolution, 127–133
on mating, 39–40, 173, 178–184
women refuting, 42–43
Sociobiology: A New Synthesis (Wilson), 36–37
sociology, 20
Socrates, 23
Spanidou, I., 85–86
spatial abilities, 77, 122
hunting's role in evolution of, 127–133
sex differences in, 119, 124–127
species, 28, 99
Spencer, H., 110
sperm, 92, 142–144
Spitzka, E. A., 34–35
status. *See* class, social
steroids, receptors for, 49
Stone-Age brain, 40, 90, 92
stress
effects in pregnancy, 151–152, 199
effects on hormone levels, 73, 151–152
"Strong Male-Driven Evolution of DNA Sequence" (Li and Makova), 143
Subjugation of Women, The (Mill), 93
susceptibilities, 15–16
Swaab, D. F., 203
symmetry, 42, 180–182

Taylor, J., 41
technology, reductionism in, 10
territoriality, 37
testosterone. *See also* hormones
in brain development, 52, 78, 149
effects on cognitive abilities, 118, 125–126, 133
effects on parental behavior, 54–55, 78
influence on sexual behavior, 52, 170–171
levels in pregnancy, 151–152
relation to estrogen, 49–50
theater, double-plays with gender in, 185–189
Theweleit, K., 115
Thorndike, E., 35
Thornhill, R., 39, 180–181
Tiger, L., 40, 111
Tocqueville, A.-C.-H. de, 89
Tooby, J., 69, 89–90, 127
toys, sex differences in, 57–60, 150–151, 153
transsexuals/transgenders, 20, 146–147, 189–190, 222

transvestites, 185–189
Trivers, R. L., 37, 111
twin studies, 216–220

United Nations Declaration of Human Rights, 98

variability, 117, 145–147, 169. *See also* flexibility
verbal abilities, 153
  brain activity during, 136–137
  on IQ tests, 122–124
Voltaire,, 23
vom Saal, F. S., 149
von Gruber, M., 115

Wahlsten, D., 77–78
Walker, A., 101
Walter, N., 95
Walton, P., 93
Wandor, M., 185–186
war, 89–90, 93. *See also* aggression
Weber, M., 38
Wedekind, C., 179
Weinberg, T. S., 152
West, D. J., 198
What a Drag (Dickens), 187
Willard, D., 37
Williams, T. J., 221
Wilson, E. O., 12–13, 36–37, 89, 111, 208
Witelson, S. F., 77
Wolf, N., 106
Wollenstonecraft, M., 85, 95
"woman," meanings and use of, 99
Woolf, V., 225

Yahr, P., 50
Young, E., 204
Young, I., 88